Foundations of Data Science and Statistics: Analytics Made Simple

Woody R. Clermont

Woody Clermont Book Publications.

Book Cover by The Unrelenting Alchemist©

Published by Woody R. Clermont

Illustrations by The Unrelenting Alchemist

First edition 2025.

Preface

I didn't write this book for math prodigies or data scientists. I wrote it for everyday people—people who may have struggled with math, doubted themselves, or been told somewhere along the line that "this just isn't for you." I'm here to tell you it is.

I have an Associate of Science in Computer Science, a Bachelor of Science in Mathematics, and Master's degrees in Business Administration, Economics, and Data Analytics. My academic journey gave me a strong foundation in logic, computation, and real-world problem solving. But my passion for teaching is what led me to create this book.

For many years, I taught both Microeconomics and Macroeconomics at Miami Dade College—an institution known as the largest college in the United States. But the course I spent the most time teaching, and perhaps the one closest to my heart, was QMB 2100: Basic Business Statistics.

In those classrooms, I met students from every walk of life. Some were fresh out of high school; others were returning after raising children or changing careers. One student to my surprise, was homeless. Some were confident; others were terrified. But each of them had a story—and a struggle. My job was to lift them all to a level of success that they never thought was possible.

I never believed in the kind of teaching that favors only the naturally gifted or rewards only those who "get it" quickly. Instead, I believe in teaching that adapts, supports, and empowers. I designed my courses to ensure that everyone could learn the material. I believe in mastery, not memorization. Growth, not gatekeeping.

That belief was tested when I was asked to teach accelerated sections of the course—classes where students had to master a full semester's worth of statistics in just eight weeks. It was tough. The pace was fast. But the success rate? Remarkably high. Why? Because my students bought in. We established clear rules of engagement. And I promised to meet them every step of the way—with clarity, encouragement, and structure.

This book reflects that same teaching philosophy. It is structured to be accessible. It offers real-life examples that matter. And most importantly, it assumes that you—yes, you—can learn this. Data literacy isn't about being gifted. It's about being guided.

To show you how to make concepts easier to graph, here is a way to conceptualize something occurring in four dimensions, which some people may struggle to imagine.

To represent a four-dimensional (4D) object or function, one effective method is to fix one dimension—say time

t—and generate a series of 3D plots for discrete values of t. Each plot provides a "slice" of the 4D space at a given time, allowing us to study the behavior of the 3D system as t evolves.

This approach mimics how movies are created: a sequence of static images (frames) displayed rapidly to simulate motion. In this case, we treat the diagonal line of increasing t as our timeline and observe how the 3D graph changes.

Let's consider the function:

$$f(x, y, z, t) = \sin(x + y + z + t)$$

We evaluate this function for three discrete values of t: -2, 0, and 2, then plot the 3D slice for a fixed z value to visualize the spatial changes in x and y and z.

Below are three 3D visualizations of the function:

$$f(x, y, z, t) = \sin(x + y + z + t)$$

evaluated at $t = -2, 0,$ and 2. Each image is a snapshot showing how the 3D spatial form of the function changes as time progresses.

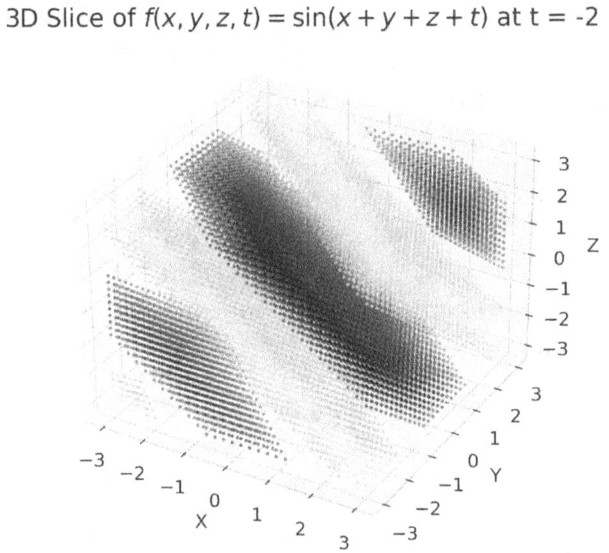

Figure 1: 3D Slice at $t = -2$ of $f(x, y, z, t) = \sin(x + y + z + t)$

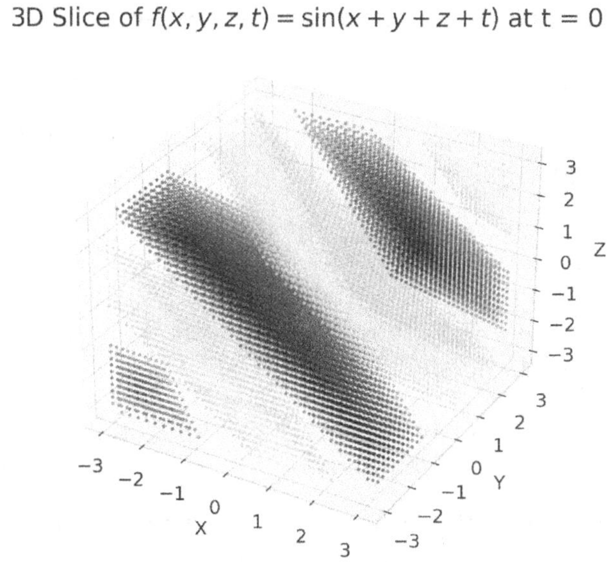

Figure 2: 3D Slice at $t = 0$

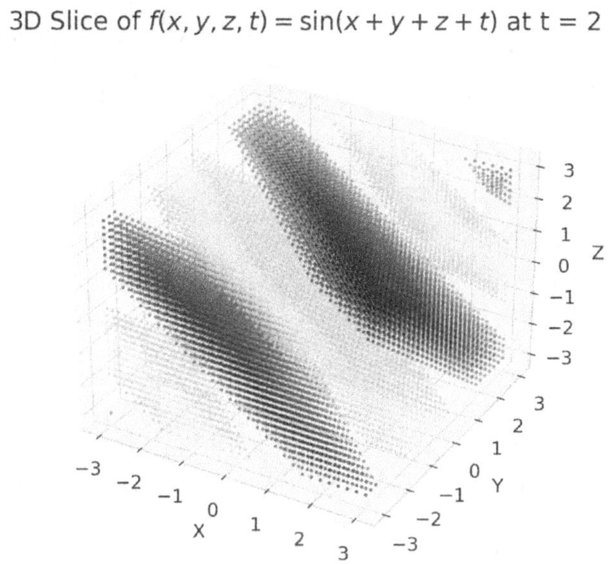

Figure 3: 3D Slice at $t = 2$

This technique is widely used in:

Physics simulations (e.g., evolving scalar/vector fields over time)

Topology (e.g., visualizing 4D polytopes like the tesseract)

Data science (e.g., how a 3D scatter plot shifts as a parameter changes)

Mathematics and **machine learning** (e.g., slicing high-dimensional manifolds)

We can then realize there are easy intuitive ways to explain even seemingly hard to understand concepts.

Welcome to the journey.

Woody R. Clermont

Contents

Chapter 1

Why Data Matters?

1.1 The Age of Data

We live in a world drowning in data. Every action you take, whether online or offline, creates a footprint. When you unlock your phone, it logs the time. When you check the weather, you're retrieving real-time measurements from thousands of sensors. When you vote, your choice becomes a tiny entry in a massive spreadsheet of civic behavior. When you click "like" on a social media post or search for the cheapest flight, you're contributing to behavioral data that platforms use to learn your preferences.

We are not just passive consumers of data. We are active producers of it. Each tap, swipe, purchase, and interaction feeds into a vast ecosystem of databases and algorithms designed to optimize your experience, or exploit it.

Big Data, **AI**, and **machine learning** might sound like buzzwords reserved for tech companies and scientists, but these systems impact everything from what job ads you see to the cost of your car insurance. If we fail to understand the fundamentals of data, we lose the ability to interpret the world accurately and protect our autonomy.

In short: **data fluency is not optional**. It's a survival skill in the digital age.

1.2 The History of Data

When you hear the word "data," you probably think of spreadsheets, computer servers, or big tech companies. But the story of data goes back thousands of years, to ancient civilizations, early writing systems, and the need to remember, record, and reason.

Etymology of "Data"

The word *data* is the plural form of the Latin word *datum*, which means "something given" or "a thing that is granted." In ancient Latin, a *datum* was a single point of information—a piece of knowledge offered up for consideration. The word "data," then, literally means "things that are given."

Datum is a neuter noun of the second Latin declension.[1]

Originally, "data" referred to established facts or principles—something accepted as a foundation. For example, a medieval mathematician might say, "Let us begin with the data," meaning "Let us start with what we already know."

In modern English, the word "data" evolved to mean raw information collected for analysis or reference. Interestingly, while "data" is technically plural, most people today use it as a singular mass noun (like "sand" or "furniture"). So you'll often hear, "The data is helpful," rather than, "The data are helpful."

Prehistoric Records

The earliest known use of data was not digital, it was visual, physical, and deeply practical. Long before written language, early humans carved *tally marks on bones and cave walls* to track objects, moon phases, or seasonal changes. These tally systems, some dating back over 30,000 years, represent humanity's first attempts to record observable quantities [4, 11].

One of the oldest examples is the *Ishango bone*, discovered in the Democratic Republic of Congo and

[1] In Latin, *-um* is the nominative, vocative, and accusative singular ending for neuter second-declension nouns. These forms do not change between subject, direct object, or noun of address. The plural form for all three cases is *-a*. Since neuter nouns refer to objects or abstract concepts (neither male nor female), they use this distinct morphological pattern. Thus, *datum* becomes *data* in the plural.

believed to be over 20,000 years old. The bone has carved notches arranged in groups, possibly representing arithmetic or calendar patterns [5, 13, 7].

As civilizations developed, data became a cornerstone of trade, governance, and agriculture:

- **Sumerians (c. 3000 BCE):** Used cuneiform writing on clay tablets to record grain counts, labor rosters, and tax payments. This was possibly the world's first bureaucracy, built on data.
- **Ancient Egyptians:** Maintained detailed records of harvest yields, census rolls, and construction supplies. The great pyramids could not have been built without managing volumes of logistical data.
- **Chinese Dynasties:** Collected population data and agricultural records to determine taxes and military service. The Han dynasty (206 BCE – 220 CE) conducted one of the earliest full censuses in history.
- **Roman Empire:** Known for organizing vast amounts of data on land ownership, census rolls, and military strength. The word "census" comes from the Latin *censere*, meaning "to assess or estimate."

Birth of Science

As societies began seeking not just to record the world but to explain it, data became the foundation of science

[8].

In the Islamic Golden Age (8th–13th centuries), scholars in Baghdad, Cairo, and Andalusia began cataloging astronomical data, medical case studies, and mathematical patterns. This legacy of empirical observation influenced later European scientists [14, 15].

During the Renaissance and Enlightenment, thinkers like Galileo, Kepler, and Newton began formalizing methods of *data collection and hypothesis testing* [6, 3]. For them, data wasn't just something to record—it was a way to test reality.

The 17th century also saw the rise of *probability theory*, developed by Pascal and Fermat. This was a shift in how data was understood: not just as records of what happened, but tools to predict what might happen [9].

Industrial Era

By the 1800s, with the rise of empires and industry, data was seen as a tool for power and efficiency. Governments and businesses began collecting massive datasets to improve logistics, taxation, labor allocation, and public health systems.

- **National censuses** became regular in many countries, guiding policy and resource allocation. Census data was used to determine everything from military enlistment to railroad construction.
- **Florence Nightingale** used hospital mortality data in the Crimean War to revolutionize

healthcare and hygiene. Her famous rose diagrams—early forms of data visualization—helped persuade British officials to implement sanitary reforms in military hospitals [16].

- **Statistical societies** were founded across Europe, such as the Statistical Society of London in 1834, promoting data as a civic and scientific tool.

At the same time, efforts were made to automate the processing of this growing volume of information. This gave rise to the concept of mechanical computing:

- **Charles Babbage**, a British mathematician and inventor, designed the *Difference Engine* in the 1820s—an automatic calculator meant to compute polynomial functions. Though never fully completed, it demonstrated that calculation could be mechanized [17].

- Later, Babbage conceptualized the *Analytical Engine*, a general-purpose computing machine with components that closely resemble modern computers: a memory unit, processor, input/output devices, and the ability to follow a sequence of instructions.

- **Ada Lovelace**, working with Babbage, created what many consider to be the first algorithm designed for a machine. She envisioned that the Analytical Engine could manipulate not just numbers, but symbols and ideas, making her the world's first computer programmer [18].

These breakthroughs foreshadowed a world where data could be handled not just by hand but by machines.

Data was now political, economic, and increasingly technological. Whoever had the best data—and the means to analyze it—had the best chance of controlling people, economies, and territory.

This era culminated in the foundational work of Alan Turing in the 20th century, who formalized the concept of the algorithm and developed the mathematical model now known as the *Turing machine*. His ideas laid the groundwork for modern computing, establishing the principle that any computable problem could be solved by a sufficiently capable machine following formal instructions [10].

Digital Revolution

In the 20th century, data exploded in quantity and speed. First with punch cards, then with electronic computers, and finally with the Internet and smartphones, data became a living, constantly expanding force.

Here are a few key moments:

- **1940s–50s:** The invention of digital computing made it possible to store and process more data than ever before.
- **1970s:** Relational databases allowed companies to organize massive information systems (e.g., customer orders, payroll).

- **1990s:** The rise of the Internet transformed how data was shared, stored, and generated.
- **2000s–present:** Big Data, artificial intelligence, machine learning, and cloud storage created an environment where billions of data points are created every second.

Today, some estimate that *90% of all human data has been created in just the last two decades.* We generate data from everything: smart watches, credit card swipes, web searches, and even refrigerators.

Why Learn This?

Understanding the history of data helps us appreciate:

- That data is ancient, not just modern.
- That data has always been tied to power—whether it was feeding armies, predicting floods, or setting interest rates.
- That our role in the data system has changed: we are not just observers, but generators and subjects of data.

The tools may be new, but the human need to measure, count, compare, and understand is timeless. Learning the history of data reveals that this literacy is not about computers, it's about consciousness.

1.3 Math is not Hard

Let's get something clear from the start: **Everyone can do math.**

This book is about patterns, meaning, and decision-making. It's about learning to recognize when numbers are being used to tell the truth—or to mislead. It's about developing a gut instinct for spotting fishy statistics or confusing graphs.

Here's what you will learn instead:

- How to spot misleading graphs on social media or in the news.
- How to ask better questions during a presentation at work or in class.
- What it actually means when someone says there's a "margin of error."
- How to make smarter everyday decisions—from choosing a cell phone plan to interpreting a poll before an election.
- When to trust a data source and when to dig deeper.

Many students think, "I've always been bad at math." But being "bad at math" often means you were never shown how it applies to your world. If you've ever compared prices, estimated how long a task might take, or decided whether to trust a review, you've already used data thinking. You just didn't call it that.

It's important to understand that everyone is capable of learning. Struggling with math—or with data—does not mean you're incapable. It often means you weren't given the right learning environment. In the 1960s, educational psychologist **Benjamin Bloom** introduced the concept of *Mastery Learning*, which challenged the idea that only some people are "naturally gifted" at learning. Instead, Bloom argued that almost all students can achieve a high level of understanding if they are given enough time, the right instructional strategies, and consistent support [1].

Educators like **John B. Carroll** and **Fred Keller** also supported this view, developing methods such as the Personalized System of Instruction to ensure learners could move at their own pace and master each concept before advancing [2, 12]. These approaches recognize that people learn in different ways and at different speeds—and that doesn't make anyone inferior. It just means education should be flexible and responsive.

So if you've ever thought, "I'm just not a numbers person," remember: that label isn't true. Your brain is wired for pattern recognition, estimation, and logical reasoning—skills that are at the heart of data literacy. With the right support and tools, you can learn this. In fact, you already have.

1.4 What Is Data Literacy?

Data literacy is the ability to read, interpret, analyze, and communicate using data. Think of it as learning a new kind of language—one that allows you to ask smarter questions, detect inconsistencies, make informed decisions, and avoid being misled.

In today's world, being able to "speak data" is just as important as being able to read and write. It allows you to navigate everything from budgeting your personal finances to understanding political polls or workplace metrics.

Not Memorization

You don't need to become a statistician or remember long equations. Data literacy is not about complicated math—it's about curiosity and critical thinking.

Here are the types of questions a data-literate person asks:

- **Where did this data come from?** Was it a reputable source like the U.S. Census Bureau, or a biased blog post?
- **Who collected it, and why?** Understanding motivations helps identify potential bias or limitations.
- **Is this chart telling the full story?** Are the axes manipulated? Is something conveniently missing?
- **What might be hidden or misrepresented?** Sometimes, what isn't shown is just as important as what is.

Data doesn't "speak for itself." It needs

interpretation—and your job is to learn how to interpret it wisely.

Pillars of Data Science

A truly data-literate person is equipped with four essential abilities:

1. **Read data:** Understand charts, tables, dashboards, and summaries.
2. **Evaluate data:** Identify sources, examine methodology, assess bias or uncertainty.
3. **Use data:** Apply it to solve problems, make decisions, or explain a viewpoint.
4. **Communicate data:** Turn insights into words, stories, presentations, and decisions that others can understand.

Misuse and Manipulation

Let's be honest: data is often used to manipulate. Headlines love numbers because they seem definitive—"9 out of 10 doctors...", "The average worker earns...", "60% of Americans agree..." But without knowing how that data was collected, how big the sample was, or what the question asked, you might be getting a skewed picture.

As a data-literate reader, you will learn to:

- **Recognize trustworthy vs. shady sources.** Is

this based on peer-reviewed research? Government data? Or a TikTok comment thread?

- **See through manipulation.** Learn to spot the tricks: selective graphs, misleading scales, cherry-picked statistics.
- **Make informed choices.** Whether you're choosing between health plans, analyzing public policy, or debating with friends, data gives your decisions more power.

1.5 Data is Everywhere

Data is everywhere—and once you start to look, you'll see it in more places than you ever expected. It's woven into nearly every modern system, product, and process.

- **Netflix Recommendations:** Based on your viewing history, pause length, genre preferences, and even the time of day you watch. It's pattern recognition, not magic.
- **News Polls:** "52% of voters support Measure A." But did they ask 100 people or 10,000? Was it online, in-person, or phone-based? Did the sample represent the whole population?
- **Weather Forecasts:** A "70% chance of rain" doesn't mean it will rain over 70% of the city, or 70% of the time. It means that, historically, under similar conditions, rain occurred 70% of the time.
- **Spotify Playlists:** Your listening behavior is

compared with thousands of others using clustering algorithms. It's like digital sociology.

- **Job Applications:** Many resumes are filtered through software. Applicant Tracking Systems (ATS) assign scores based on keywords, years of experience, and degrees. If you're unaware of how this data-driven system works, your resume may never even be seen by a human.
- **Healthcare and Public Health:** Data is used to detect disease outbreaks, allocate medical resources, and tailor health campaigns based on community need.

Systems Around You

Once you begin thinking in terms of data, you start noticing systems. You realize every store has point-of-sale data. Every public bus uses ridership metrics. Every digital advertisement is based on your past behavior and someone else's forecast. This awareness is empowering—it helps you regain control of how you interact with systems that once operated invisibly.

1.6 Census Table

Let's look at a simplified example of real data. The U.S. Census Bureau regularly publishes datasets on demographics, income, education, and housing. Below is a sample excerpt inspired by the American Community

Survey (ACS) data.

Table 1.1 – Demographic Snapshot: Miami-Dade County, Florida (2022)

Category	Population	% Hispanic	Median Age	Poverty Rate
Miami-Dade County	2,673,837	69.4%	40.2	15.2%
Florida (Statewide)	22,244,823	27.1%	42.4	12.7%
United States (Total)	332,403,650	18.7%	38.5	11.6%

How to Read

Each row represents a geographic level: county, state, and national. Each column gives a different statistic about that region's people.

What You Can Learn:

- Miami-Dade has a much higher percentage of Hispanic residents (69.4%) compared to the U.S. average (18.7%). That cultural difference might affect language services, voting materials, and education needs.
- The median age in Miami-Dade is slightly younger than the state average, but slightly older than the national average.
- The poverty rate is higher than both Florida and the national average. This could influence housing programs, social services, and funding priorities.

This one table can inspire dozens of questions and insights: How has this changed over time? What policies are in place to address disparities? Are trends improving or worsening?

Key Lesson

When you read a data table, you're reading a narrative. In this case, the story is about demographics and inequality across geographic levels. A well-designed data table is like a paragraph—it's compact, full of meaning, and requires interpretation.

Data literacy gives you the ability to read that paragraph and explain it to others.

1.7 See Data?

Data is hiding in plain sight. You don't need to be a scientist or statistician to realize you're surrounded by it. Let's break down where you encounter data in everyday life:

- **Netflix Recommendations:** Every watch, pause, skip, or binge contributes to an algorithm that tailors content just for you. But it also reveals broader trends in human behavior.
- **News Polls:** "52% of voters say..." is based on sampled data. Understanding what that means—how the sample was collected, how many people were surveyed, and what the margin of error is—makes you a critical reader.
- **Weather Forecasts:** These are probability

models based on historical data and current measurements. When they say there's a "70% chance of rain," that means that in similar conditions, it rained 70% of the time.

- **Spotify Playlists:** Your taste in music is compared against thousands of users to create curated playlists. This is clustering and recommendation data science in action.
- **Job Applications:** Applicant Tracking Systems (ATS) use data to sort resumes. Keywords, degrees, and experience are filtered by automated systems long before a human ever sees your application.

You don't need to learn to code. You just need to become more aware of how these systems work—and how you're already part of them.

1.8 Applications

You might think that data is only relevant to people in tech, science, or business. But the truth is, data literacy is crucial across every field and every profession. Here are some examples:

- **Social workers** use demographic data to understand community needs, analyze case trends, and report outcomes to agencies.
- **Artists and creatives** look at audience engagement data to fund, promote, and grow their work. Art exhibitions use foot traffic data;

YouTube and TikTok rely on viewer retention metrics.

- **Journalists** use statistics from public records, polls, and research studies. A good journalist knows how to read a dataset and how not to be misled by one.
- **Psychologists** collect and interpret data in experimental design, surveys, and outcome measurements.
- **Activists and policy advocates** rely on data to show social trends, racial disparities, health outcomes, or incarceration rates. Facts make a stronger case than anecdotes.
- **Teachers** analyze student performance data to improve learning strategies. Even lesson planning can benefit from tracking what works and what doesn't.
- **Retail workers** and small business owners review sales data, customer feedback, and product returns to make better inventory decisions.

In short: **every career is becoming data-driven**. The better you understand data, the more empowered, employable, and effective you become.

1.9 Will Help You

This book will not turn you into a data scientist. That's not the goal. Instead, this book will give you:

1. A toolkit to decode graphs, charts, polls, and figures with confidence.
2. Real-world examples of how data is used (and misused) in your daily life.
3. Critical thinking exercises that show you how to question and analyze claims.
4. Hands-on practice using Excel, Google Sheets, and free tools to explore datasets.
5. A deeper awareness of how your actions generate data and what that means for your privacy and decisions.

You'll encounter:

- **Case studies** showing how students, artists, and activists use data in real life
- **Myth-busting boxes** that challenge popular misconceptions (like "the average is always typical")
- **Practice problems** that feel more like puzzles than math
- **Quick tips** for everyday decisions, like comparing credit cards or understanding medical test results

And most importantly, you'll finish each chapter more confident in your ability to handle the truth—when it's wrapped in numbers.

Quick Review

- We live in a data-driven world where understanding data is essential for participation and protection.
- You don't need to be "good at math" to become data literate.
- Data literacy is about questioning, interpreting, and applying evidence.
- This book is your bridge from data confusion to data clarity—no jargon, no fear.

Reflection Questions

Understanding data literacy is more than just absorbing facts—it's about learning how to reflect critically on how numbers, charts, and patterns influence your daily life. Use the following expanded questions to explore your personal relationship with data, your learning goals, and your future use of this knowledge.

1. **What is one area of your life where you now realize data plays a role, even though you never noticed it before?** Consider your shopping habits, your social media feed, or even your favorite streaming service. What kinds of data are being collected? How do you think that data shapes what you see or do next? Would you change your behavior if you were more aware of it?

2. **Can you remember a time when someone used numbers to try to convince you of something?** This could be in an advertisement, a news article, a political campaign, or a classroom discussion. Were you persuaded? Looking back, do you feel those numbers were fairly used, or do you now see that something might have been left out or exaggerated?

3. **Think of your future career. How do you think data literacy might give you an edge over others who don't have it?** Be specific: Will you use data to manage people, track progress, support a cause, write more effective reports, or avoid being manipulated? How would a strong understanding of data help you stand out in your field or earn trust from others?

4. **What are you most afraid of when it comes to learning about data?** Be honest. Is it the math? The terminology? The feeling that it's "not for people like me"? Write out your fears. Then, name one thing that would help reduce that fear: a visual guide, a good teacher, relatable examples, or maybe seeing someone like you succeed at it.

5. **Describe a time when you made a decision—big or small—without having full information.** Would data have helped? For example, did you buy something without reading reviews, or vote without looking at polls or background facts? What might you have done differently if you'd had clearer data?

6. **Think about the last chart or graph you saw.** Where was it—from a news outlet, social media, class, or a product label? Did it make sense to you? Were the labels clear? How might the same data have been presented in a way that was more helpful?

7. **Do you trust the data you see online? Why or why not?** List examples of trustworthy and untrustworthy sources you've encountered. What makes one feel more reliable than another? Would you feel confident explaining to someone else how to spot a biased chart or misleading statistic?

8. **What would your "data diary" look like for one day?** Track how many times in a single day you interact with data: checking the weather, looking at your bank balance, reading reviews, interpreting a calorie count, or seeing election poll numbers. How many decisions do you make based on data—consciously or not?

9. **If you were designing a course for high school or college students on data literacy, what would you include?** Would you focus more on interpreting media, analyzing statistics in sports, understanding health trends, or practicing financial literacy? What's the most important thing you think people your age need to understand?

10. **How might your relationship with data change after reading this book?** Try to imagine how you'll think differently a month from

now. Will you feel more confident? More skeptical? Will you pause before reposting a statistic, or challenge a misleading graph? Describe how you hope this journey will transform you intellectually and practically.

Chapter 2

Types of Data

Before we can analyze data, we need to know what kind of data we're working with. Just like sorting laundry before doing a wash—whites, colors, delicates—data needs to be sorted too. The way we handle and interpret it depends on its type.

In this chapter, we'll explore the main categories of data: categorical and numerical, structured and unstructured. You'll learn how to recognize them in the wild and how they behave differently when you analyze them.

2.1 Categorical vs. Numerical

At the most basic level, data can be sorted into two primary types: **categorical** and **numerical**. You may also hear these referred to as **qualitative** and

quantitative data, respectively.

- **Qualitative (categorical) data** describes labels, names, types, or groupings. It focuses on classification rather than measurement.
- **Quantitative (numerical) data** represents measurable quantities and involves values that have mathematical meaning.

Understanding this distinction is foundational because it influences what kind of graphs you use, what statistical tools are appropriate, and what conclusions can be drawn from your data. Misclassifying data can lead to misleading visualizations or incorrect interpretations.

For instance, you would never average a list of ZIP codes (categorical), but you could average a list of incomes (numerical). Treating categorical data like it's numerical (or vice versa) is one of the most common early mistakes in data analysis.

Categorical Data

Categorical (qualitative) data represents *groups*, *labels*, or *names*. It answers the question: **"What type?" or "What kind?"**

It's important to note that categorical data may consist of words or numbers. For example, student ID numbers, even though they're digits, are *categorical*, because they don't represent a measurable quantity—you can't average them or do arithmetic on them.

- **Eye color:** blue, green, brown

- **Car brand:** Toyota, Ford, BMW
- **Class level:** freshman, sophomore, junior, senior
- **Survey responses:** yes, no, maybe
- **T-shirt size:** small, medium, large, extra-large
- **Marital status:** single, married, divorced, widowed

Binary vs. Multi-Level Categories:

- **Binary data** has only two possible categories (e.g., yes/no, true/false, pass/fail).
- **Multi-level categorical data** includes more than two possible labels (e.g., types of cuisine: Italian, Chinese, Mexican, Thai).

Two Subtypes of Categorical Data:

- **Nominal data** has no inherent order or ranking. Examples: gender, eye color, language spoken.
- **Ordinal data** includes categories with a logical order, but the intervals between them may not be equal. Examples: satisfaction rating (poor, fair, good, excellent), education level (high school, bachelor's, master's, doctorate).

Real-Life Analogy: Think of categorical data like clothing tags. You might see tags for size (S, M, L), brand (Nike, Adidas), and color (red, blue, black). These labels don't tell you "how much" of anything, but they help you organize your wardrobe—and make buying decisions.

Caution: Just because a data value is written as a number doesn't mean it's numerical. For example, ZIP codes, phone numbers, and jersey numbers are all categorical—they identify, but don't measure.

Numerical Data

Numerical (quantitative) data consists of numbers that represent measurable quantities. It answers the questions: **"How much?" "How many?" or "How far?"**

This type of data is ideal for most statistical calculations, including finding means, medians, standard deviations, and performing regression analysis.

- **Age:** 18, 32, 45
- **Income:** $25,000, $65,000, $120,000
- **Temperature:** 72°F, 98.6°F, 110°F
- **Height:** 5'4", 6'1"
- **Time spent on a task:** 5 minutes, 90 minutes

Two Subtypes of Numerical Data:

- **Discrete data:** Consists of countable values, often whole numbers. You can enumerate each possible value, and gaps exist between them.

  - Example: number of children, number of courses taken, number of text messages sent today

- **Continuous data:** Can take on any value within a given range. These values are measurable, not countable, and can include fractions and decimals.
  - Example: weight, height, time, speed, distance traveled

Real-Life Analogy: Imagine a gas pump. As it fills your tank, the numbers tick upward in fractions of a cent and gallon. That's continuous data in action—it flows. Contrast this with the number of cars in the parking lot—that's discrete data. You can't have 3.6 cars.

Operations You Can Perform:

- You can calculate the mean (average), median, mode, range, and standard deviation.
- You can graph it using histograms, line graphs, box plots, and scatterplots.
- You can use numerical data to make predictions, correlations, and comparisons.

Why This Matters

Understanding whether your data is categorical or numerical is like knowing the difference between ingredients when cooking. If you confuse sugar with salt, or baking soda with baking powder, the outcome of your dish could be inedible. Similarly, if you mistake a category label for a countable number—or try to compute an average of something that can't be measured—you may end up with conclusions that are misleading, inaccurate, or completely invalid.

Imagine you're a small business owner tracking customer feedback. If you treat ratings like "excellent," "good," "fair," and "poor" as though they were evenly spaced numbers, you might calculate an "average rating" that doesn't really reflect what your customers are saying. Worse yet, if you try to perform calculations on ZIP codes or Social Security numbers—thinking they're numerical—you'll end up with results that have no real meaning. The consequences of misunderstanding data types can range from embarrassing errors in a presentation to making flawed business or policy decisions.

Here's why this distinction is foundational:

1. The Right Graph for the Right Data
Visualizing data without knowing what type it is often leads to miscommunication. For categorical data, you'll typically use bar charts, pie charts, or stacked column charts—these show proportions and counts. But if you tried to use a histogram for something like shoe brands, it would confuse your audience. On the other hand, for numerical data, histograms, box plots, scatter plots, and line graphs reveal patterns in magnitude, spread, and trends over time. Choosing the wrong graph isn't just a stylistic mistake—it misrepresents reality.

2. The Right Statistics for the Right Data Categorical data isn't designed for calculations like averages or standard deviations. Trying to take the "mean" of favorite pizza toppings makes no sense. But numerical data begs for mathematical tools—you can compute a mean salary, a median age, or the standard deviation of travel times. Confusing the two leads to meaningless math.

Even within numerical data, whether something is discrete or continuous affects which statistical tests you can use. Statistical software requires you to classify your variables correctly before running an analysis. Get this wrong, and your p-values, correlations, or regression models won't be valid.

3. The Right Questions Lead to the Right Insights Categorical and numerical data help you answer different types of questions:

- Categorical: What type? Which group? How often?
- Numerical: How much? How many? What is the average?

For example, if you're trying to improve a public park, categorical data might tell you which activities are most popular (e.g., walking, biking, picnicking), while numerical data might reveal the average time visitors spend or the number of steps taken per visit. Both types of data provide valuable but distinct insights.

4. Accurate Communication Builds Trust In the real world, especially in business, journalism, healthcare, and politics, decisions are often based on communicated data. Your ability to describe your findings in an accurate and meaningful way depends entirely on your ability to classify the data correctly. If your audience loses trust in your understanding of the data—or senses that you're fudging numbers—they may doubt everything you say.

5. Informed Decisions Require Correct Foundations Doctors rely on numerical data (like blood pressure or glucose levels) to make diagnoses, but they also depend on categorical data (like blood type or diagnosis category) for treatment decisions. Retailers use categorical data to segment customers (e.g., new vs. returning) and numerical data to forecast demand or optimize inventory. If you build your conclusions on the wrong data foundation, your decisions—no matter how sophisticated they seem—may collapse like a house of cards.

6. Machine Learning and Automation Demand Precision In today's world of AI, automation, and predictive modeling, feeding data into systems without proper classification can be dangerous. Computers cannot interpret context like humans. If you code categorical variables incorrectly, your algorithms may create nonsensical relationships and biased outputs. This is why data scientists spend so much time on "data cleaning" and "feature engineering"—tasks that begin with understanding data types.

7. It Shapes Your Thinking Beyond technical correctness, this distinction helps shape your way of thinking. When you start looking at the world through the lens of "What kind of data is this?", you become more analytical, more precise, and more curious. You'll stop accepting every chart or statistic at face value. You'll start asking smarter questions and making better decisions—whether in your personal life, your career, or your civic engagement.

8. The Big Picture: From Theory to Action Many social problems today—like inequality, healthcare access, or misinformation—require a society that can think clearly about data. Categorical vs. numerical isn't just an academic distinction. It's the first step toward data literacy, which is increasingly a form of power. It determines who gets heard, who makes decisions, and who gets left behind.

Bottom line: Always start by asking yourself, "What kind of data am I working with?" This simple habit will unlock deeper understanding, prevent costly mistakes, and make you a more empowered thinker in every area of life.

Matching Exercise

Match each real-world example to the correct type of data and measurement scale. Write the letter of the correct category and the number of the correct scale next to each item.

Example	Data Type (A/B)	Scale (1–4)
1. Customer satisfaction rating: excellent, good, fair, poor		
2. Shoe size (e.g., 9.5, 10)		
3. Eye color: green, blue, brown		
4. Temperature in Celsius		
5. Zip code of a respondent		
6. Number of books read last month		
7. Time spent on social media (in minutes)		

Data Types:

A. Categorical (Qualitative)

B. Numerical (Quantitative)

Measurement:

1. Nominal
2. Ordinal
3. Interval
4. Ratio

Matching Exercise

Match each real-world example to the correct type of data and measurement scale. Write the letter of the correct category and the number of the correct scale next to each item.

Example	**Type (A/B)**	**Scale (1–4)**
Eye color (e.g., green, blue)		
Annual income in dollars		
Customer satisfaction: poor, fair, good, excellent		
Temperature in Fahrenheit		
Height in centimeters		
Number of siblings		
Car brands (Toyota, Honda, Ford)		
Rank in a competition (1st, 2nd, 3rd)		

Quick Reference Table

Category	Categorical (Qualitative)	Numerical (Quantitative)
Description	Data representing groups or categories	Data representing counts or measurements
Subtypes	Nominal, Ordinal	Discrete, Continuous
Example Variables	Hair color, brand, zip code, Likert scale	Age, income, GPA, miles per gallon
Graphs to Use	Bar chart, pie chart	Histogram, box plot, scatter plot
Typical Questions	What kind? Which group? How often?	How much? How many? What's the average?

2.2 Structured vs. Unstructured

Not all data comes in neat little boxes. Some data is well-organized and easy to analyze. Other data is messy, chaotic, and full of nuance.

Structured Data

Structured data is clean, organized, and fits into rows and columns—like a spreadsheet or database table.

- Student ID, name, GPA
- Product name, price, inventory count
- Date, location, temperature

Structured data is easy for computers to read and analyze. It's used in traditional statistics, accounting software, and relational databases.

Unstructured Data

Unstructured data doesn't fit neatly into tables. It's messy, complex, and often includes human language, images, or emotions.

- Text messages
- Customer reviews
- Social media posts
- YouTube videos
- Photos and audio files

Analyzing unstructured data often requires advanced tools like machine learning, natural language processing (NLP), or sentiment analysis. Despite the challenge, unstructured data contains rich information—especially about human behavior.

2.3 Scales

Scales of measurement, also known as **levels of measurement**, classify how variables are categorized and analyzed. These scales help define what mathematical and statistical operations make sense for a given type of data.

There are four main scales: **nominal, ordinal, interval, and ratio**. Each builds upon the last, offering more sophisticated ways to analyze and interpret information.

1. Nominal Scale

The nominal scale is the most basic. It categorizes data without implying any order or ranking.

- Examples: blood type (A, B, AB, O), types of pets (dog, cat, fish), gender, race
- Key traits: categories only; no natural order; cannot perform arithmetic

Use this when: you're simply naming or labeling data.

2. Ordinal Scale

Ordinal data introduces order or ranking, but the distances between values are not defined or consistent.

- Examples: class standing (freshman, sophomore, junior), rating scales (poor, fair, good, excellent)

- Key traits: ranked categories; unequal spacing; limited math operations

Use this when: the order matters, but exact differences don't.

3. Interval Scale

Interval scales maintain equal spacing between values but lack a true zero point.

- Examples: temperature in Celsius or Fahrenheit, calendar years
- Key traits: meaningful intervals; no true zero; can add and subtract but not multiply/divide meaningfully

Use this when: differences matter, but ratios don't.

4. Ratio Scale

Ratio scales include all the properties of interval scales, but with a meaningful zero, allowing for full arithmetic operations.

- Examples: weight, age, income, distance, speed
- Key traits: equal intervals; true zero; can perform all math operations

Use this when: both differences and ratios are meaningful.

Understanding these scales helps you:

- Choose the right statistical methods
- Select appropriate graphs and visualizations
- Avoid mathematical errors (e.g., calculating an average on nominal data)

In upcoming chapters, we'll explore how these scales influence the way we graph, summarize, and analyze our data—and why skipping this step can lead to misleading or incorrect conclusions.

2.4 Real-Life Analogies

Let's make this real. Below are examples from everyday platforms to show how these data types appear in life:

Dating Apps

- **Categorical:** Gender identity, relationship preference, location
- **Numerical:** Age, distance in miles, number of likes
- **Structured:** User profiles, filters, swipe data
- **Unstructured:** Bio descriptions, uploaded photos, chat messages

Apps use structured data to sort potential matches, and unstructured data to create nuanced algorithms that learn what you like.

Spotify

- **Categorical:** Genre, artist name, subscription type
- **Numerical:** Number of plays, duration in minutes, rating (likes)
- **Structured:** Track ID, playlist name, release date
- **Unstructured:** Lyrics, user comments, song reviews

Spotify uses structured data for basic functionality and unstructured data to personalize your music recommendations.

Amazon

- **Categorical:** Product category, shipping method, seller region
- **Numerical:** Price, star rating, number of reviews
- **Structured:** Item ID, SKU, order date
- **Unstructured:** Product descriptions, customer reviews, Q&A sections

Amazon's recommendation engine relies heavily on structured data to suggest products and unstructured data to assess customer satisfaction.

2.5 Why It Matters

When you understand what kind of data you're working with, you avoid using the wrong tool for the job. You wouldn't use a pie chart for continuous numerical data, or calculate a mean for qualitative opinions. Knowing the type of data guides your analysis, improves your accuracy, and helps you interpret results more meaningfully.

But this isn't just about graphs or statistical formulas. This is about **how we see the world**. Think of a medical researcher who has to distinguish between symptoms (categorical) and blood pressure readings (numerical). Or a business owner who analyzes customer feedback (qualitative) and compares sales revenue (quantitative). Without clarity about data type, conclusions become muddy, and decisions become risky.

Misunderstanding Data

When data is misunderstood, decisions can go horribly wrong.

- A policymaker might rely on average income data (a mean) in a community where income distribution is highly skewed, leading to poor allocation of resources.
- A healthcare professional might apply the wrong statistical test to patient data, misunderstanding which treatments are effective.

- A marketer might misinterpret consumer behavior by treating "yes/no" survey responses as if they had measurable distance between them.

These mistakes don't just affect graphs—they affect *people's lives, money, and well-being.*

Clarity = Confidence

When you understand the difference between categorical and numerical data—and the measurement scales like nominal, ordinal, interval, and ratio—you gain confidence. You no longer feel intimidated by a spreadsheet or overwhelmed by a data dashboard. You can ask sharper questions, make cleaner arguments, and critique flawed reasoning. You become **data literate**.

Not Just Numbers

A common myth is that data means numbers. But data can also mean *labels, choices, preferences, emotions*, and other forms of representation. Every time you click "like" on a post, swipe left or right on an app, or choose a brand at the store, you generate data. And the organizations collecting that data are often *very aware* of the distinctions—using different models to analyze your behavior depending on whether it's structured or unstructured, qualitative or quantitative.

This Chapter

Think of this chapter as the decoder ring for the rest of the book. It gives you the terminology, categories, and mental structure to look at data with clarity. Without this foundation, everything else in statistics becomes confusing, frustrating, and random. But with it, you can move forward with confidence. You're not just memorizing definitions—you're learning how to ask the right questions.

Visualizing Reality

In the chapters ahead, we'll transform these abstract distinctions into real visuals and real applications. You'll learn which types of graphs match which types of data. You'll build tables, charts, and summaries that *make sense.* You'll see how recognizing data types simplifies everything from designing a survey to pitching an idea.

Bottom line: Don't skip over data types. This is the blueprint for the entire house. And you're not just building any house—you're building one where every insight is built on solid ground.

Chapter 3

Charts & Graphs

Visualizing Numbers

Data is powerful—but raw numbers alone are rarely persuasive. Our brains process visuals far more efficiently than text or numbers. That's why charts and graphs are everywhere: in news articles, business reports, sports broadcasts, medical studies, and even social media posts.

This chapter introduces the most common types of data visualizations, explaining when and how to use them. You'll also learn to be a critical observer—not just of what a chart shows, but what it *doesn't* show. Because just like words can be twisted, graphs can be manipulated.

3.1 Bar Charts

Bar charts are one of the most commonly used visual tools in data analysis. They provide a simple yet powerful way to compare categories at a glance. Each bar represents a category, and the height (in vertical bar charts) or length (in horizontal bar charts) represents the frequency, count, or value associated with that category.

- **Used for:** Categorical (qualitative) data
- **Best for:** Displaying comparisons among discrete groups
- **Common axes:** Categories on the x-axis; frequency or value on the y-axis
- **Orientation:** Can be vertical or horizontal depending on label length or design
- **Key tip:** Bars should be equally spaced and not touch each other; width should be consistent

Bar charts shine when you want to answer questions like:

- Which category is the most popular?
- How do several groups compare with one another?
- Has a certain category grown or shrunk over time?

Real-World Analogy: Voting Results

Imagine a student government election with five candidates. A bar chart instantly shows who received the most votes and how everyone else compares. This is far easier to grasp than reading a long table of numbers.

Design Matters

Although bar charts are easy to understand, how they are designed makes a big difference. Consider the following guidelines for creating effective bar charts:

- **Axis clarity:** Always label both axes clearly. The category labels should be readable and unambiguous.
- **Consistent spacing:** Ensure equal spacing between bars. Irregular spacing can create the illusion that some categories are more related or important than others.
- **Order categories thoughtfully:** Alphabetical order is fine, but ordering by magnitude can reveal trends more effectively.
- **Avoid 3D effects:** While 3D visuals may look impressive, they can distort perception and mislead viewers.
- **Color (or grayscale) use:** Keep colors simple and high contrast. For print, use grayscale patterns or shades for accessibility.

Horizontal vs. Vertical Bars

Use vertical bars when category names are short and easily legible across the x-axis. If categories are long (e.g., "International Relations and Global Studies"), horizontal bars can make the chart more readable.

Grouped and Stacked Bar Charts

There are also variations of basic bar charts that let you add extra dimensions of comparison:

- **Grouped bar chart:** Shows subcategories side-by-side within each main category. Useful for comparing multiple variables.
- **Stacked bar chart:** Stacks values on top of one another within a single bar. Great for showing parts of a whole over time or by group.

Common Mistakes to Avoid

- **Inconsistent scales:** If your y-axis starts at a number other than zero, it can exaggerate differences between categories.
- **Too many categories:** More than 8–10 bars can make the chart cluttered. Consider summarizing or grouping smaller categories.
- **Missing labels:** A bar chart with unlabeled axes or unclear bar descriptions is almost useless.
- **Decorative but misleading visuals:** Avoid

unnecessary gradients, shadows, and animation. Focus on clear communication.

When Not to Use Bar Charts

Don't use a bar chart when:

- Your data is continuous and best suited for a histogram or line chart.
- You're working with proportions that should sum to 100
- There's a temporal trend—line graphs may better show progression over time.

Why Bar Charts Matter in Everyday Life

From news articles and business reports to school surveys and marketing dashboards, bar charts are everywhere. They simplify decision-making. A manager might use a bar chart to decide which product line needs more advertising. A student might use it to compare GPA trends across semesters. Recognizing and interpreting these charts correctly is a fundamental data literacy skill.

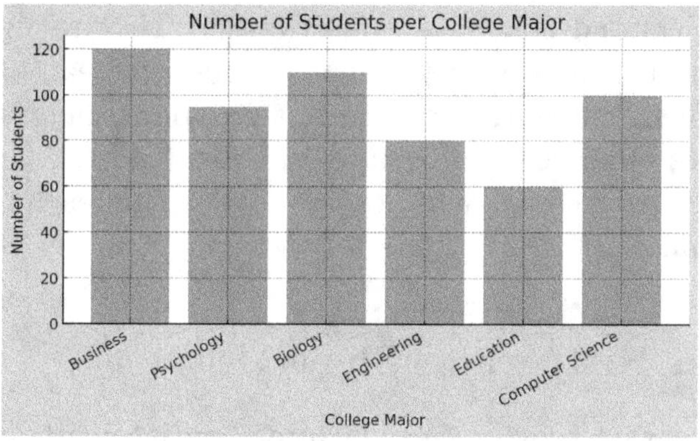

Figure 3.1 — Students by College Major
A bar chart showing the number of students enrolled in each major. This visual comparison helps identify popular fields of study and trends in academic interest.

Reflection Questions

- Can you think of a bar chart you've seen recently—in the news, at work, or on social media?
- Was it labeled clearly, or did it mislead in some way?
- If you had to explain that chart to someone else, how would you do it?

3.2 Histograms

Histograms are foundational tools for exploring and visualizing numerical data, particularly when we want to understand how data is distributed. They resemble bar charts but serve a different purpose: instead of comparing categories, histograms group continuous or discrete numerical values into intervals (called *bins*) and display the frequency of observations within each interval.

What Histograms Show Us

A histogram provides insight into the **distribution** of a dataset—its shape, center, spread, and outliers. When you look at a histogram, you're seeing how values are concentrated, whether there's symmetry or skewness, and how spread out the data is.

- **Used for:** Numerical data, especially *continuous* variables (e.g., height, weight, test scores)
- **Key tip:** Bars *should touch* because the intervals are adjacent and cover the entire number line without gaps
- **Each bar (bin):** Represents the number of values falling within a specific range

Why Bins Matter

The number and width of the bins significantly impact the appearance of a histogram:

- **Too few bins:** The graph may oversimplify and hide important details.

- **Too many bins:** The graph may appear noisy and hard to interpret.
- **Optimal bins:** Offer a balance—enough resolution to detect patterns without overwhelming detail.

For example, consider student test scores ranging from 50 to 100. If we choose 5 bins (10-point intervals), we might see a general trend. But with 10 bins (5-point intervals), we could detect nuances like a cluster near 90 or a dip around 70.

Histogram Shapes and What They Mean

- **Symmetric (Bell-shaped):** Data is evenly distributed around the center (e.g., normal distribution).
- **Skewed Right (Positive Skew):** Tail is longer on the right; many values are low (e.g., income distribution).
- **Skewed Left (Negative Skew):** Tail is longer on the left; many values are high.
- **Uniform:** All bins are approximately the same height; no concentration.
- **Bimodal or Multimodal:** Two or more peaks suggest two subgroups within the data.

Understanding these shapes can help guide your analysis. For instance, a right-skewed distribution might indicate that a few very high values are inflating the mean, making the median a better measure of center.

Histogram Example

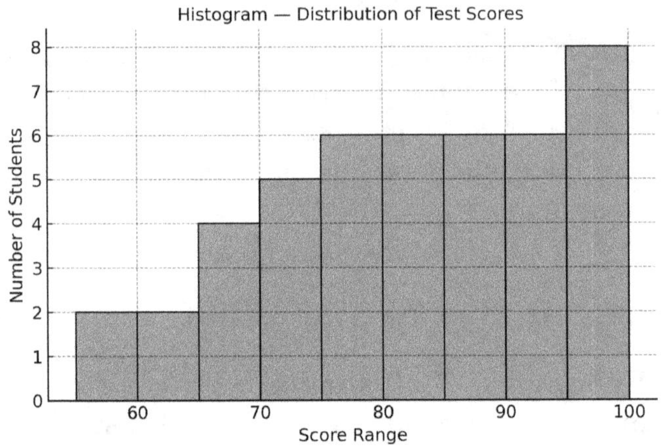

Figure 3.1: Histogram — Distribution of Test Scores

This histogram shows the distribution of test scores in a class. Most scores fall between 70 and 90, with a peak around the mid-80s. Notice how the bars are touching—this is because test scores are continuous values, and each interval directly connects to the next.

This visualization helps identify:

- **Where students performed best** (the mode)
- **If the data is skewed** or has outliers
- **Whether most students passed or failed**

Real-World Uses of Histograms

Histograms are used in a wide range of fields to summarize and explore data:

- **Healthcare:** Showing the distribution of patient ages in a clinic.
- **Education:** Analyzing score distributions to adjust teaching strategies.
- **Business:** Understanding the frequency of transaction amounts or delivery times.
- **Manufacturing:** Monitoring product measurements for quality control.

Common Mistakes to Avoid

- **Incorrect bin width:** Too wide hides variation; too narrow creates false patterns.
- **Gaps between bars:** Unlike bar charts, histograms should have touching bars.
- **Using with categorical data:** Histograms are not meant for names or labels—bar charts should be used for that.
- **Unlabeled axes:** Always label your axes with variable names and units.

Final Thoughts

Histograms are simple but powerful tools. They tell the story of your data in shape and frequency, letting you see what might otherwise be hidden in a list of numbers. Before applying statistical formulas, plotting a histogram gives you a visual pulse of your dataset.

Always ask yourself: *Does this distribution make sense for what I expected? Are there surprises?*

The histogram is often your first glimpse into the deeper story your data wants to tell.

3.3 Pie Charts

Pie charts are circular graphs divided into slices to illustrate numerical proportions. The entire circle represents a whole—typically 100%—and each individual slice corresponds to a category's relative contribution to the total. They are often used in business, media, and academic settings due to their intuitive and familiar format.

- **Used for:** Categorical (qualitative) data where proportions or percentages are important
- **Example:** Market share by brand, budget allocation, demographic composition
- **Key tip:** Use sparingly—avoid using too many categories, as this can clutter the chart and make it unreadable

Although pie charts are visually appealing and easy to

grasp at a glance, they have significant limitations. The human eye struggles to accurately judge angles, especially when comparing slices that are close in size. Unlike bar charts, which allow direct visual comparison of lengths, pie charts rely on interpreting arc angles and areas, which are less precise for analysis.

Another drawback is that pie charts do not easily support comparisons between multiple datasets. For example, if you want to compare student major distributions across several universities, bar charts or stacked column charts would be more suitable. Pie charts are best reserved for single datasets with a small number of categories.

To enhance readability, always include clear labels and percentage values on each slice. Limit the number of slices to no more than five or six. When necessary, group smaller categories into an "Other" slice to maintain clarity. Exploded pie charts—where slices are pulled out slightly from the center—can be used to highlight a particular category, but should be used sparingly to avoid visual gimmicks.

Best practices:

- Sort slices from largest to smallest starting at the top (12 o'clock position) and moving clockwise.
- Label each slice with the category name and percentage.
- Avoid 3D effects, as they distort size perception and mislead the reader.
- Keep colors distinct, especially in grayscale

printing or for color-blind readers.

Pie Chart Example: Student Distribution by Major

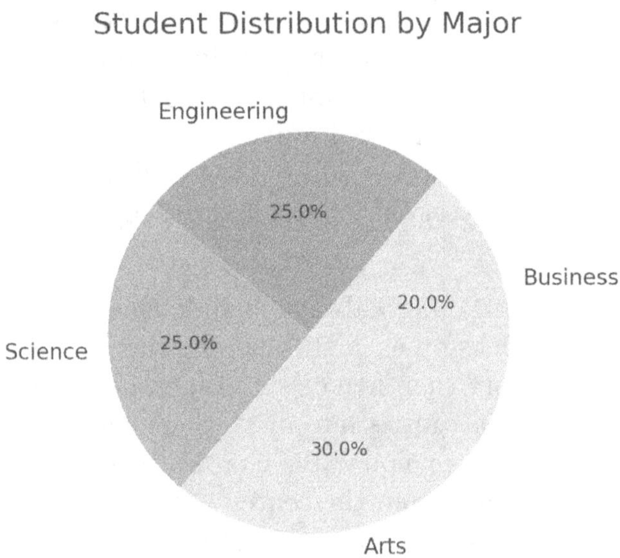

Figure 3.2: Pie Chart — Distribution of Students by Major

Pie charts are used to show how a whole is divided into parts. Each slice represents a category's proportion. In the example above, we see how students are distributed across different college majors. The chart segments the full student population into visible, labeled portions, each denoting a major like Business, Biology, Psychology, etc.

While pie charts are ideal for showing relative size and illustrating how different categories compare within a single whole, they are not good for identifying subtle differences between groups. When more precision is needed—such as comparing two groups that differ by just a few percent—a bar chart is preferable.

In summary: Pie charts are best used when:

- You need to show composition (part-to-whole relationships)
- You have a small number of categories (ideally 3–5)
- You want to provide a visual summary rather than detailed analysis

As with all data visualizations, clarity, accuracy, and honesty are more important than flashiness. If a pie chart makes your data harder to interpret—or if it looks beautiful but confuses the reader—it's the wrong tool for the job.

3.4 Line Graphs

Line graphs are one of the most effective ways to visualize how something changes over time. They connect a sequence of data points with lines to help us see trends, cycles, and shifts. Each point on the graph represents a value at a specific time, and the connecting lines tell the story of movement—whether upward, downward, or fluctuating.

- **Used for:** Time-series data, where the x-axis represents a continuous or ordered time sequence (e.g., days, months, years)
- **Example:** College enrollment trends from 2015 to 2024
- **Key tip:** Clearly label both axes, and ensure consistent time intervals to maintain accuracy and avoid confusion

Line graphs are especially useful when you're tracking progress or performance across time. They're frequently used in economics, business, health, and education to show changes in data like inflation, unemployment, temperature, sales, or academic scores. A steep slope suggests rapid change, while a flat line implies stability.

They can also show:

- **Seasonality:** Patterns that repeat in cycles (e.g., higher sales during holidays)
- **Volatility:** How much a value fluctuates up and down
- **Momentum:** Whether trends appear to be accelerating or slowing down
- **Forecasting potential:** When combined with trend lines or regression analysis, line graphs can help predict future outcomes

But be cautious—line graphs can be misleading when the y-axis is manipulated. Compressing or stretching the axis can exaggerate or hide real changes. For example, a change from 100 to 110 might look dramatic if the y-axis starts at 95 instead of 0.

Line Graph Example: College Enrollment Over Time

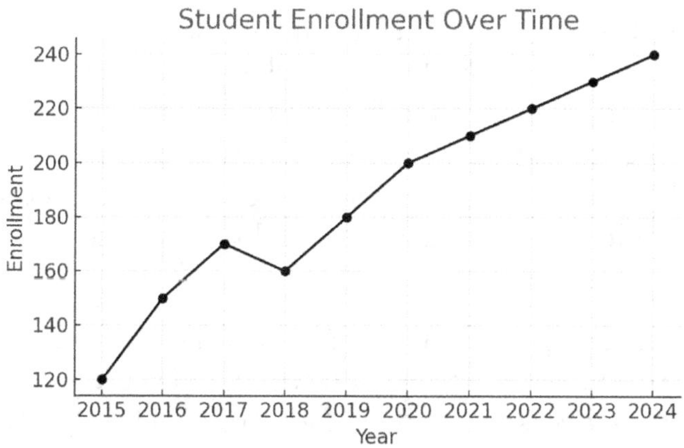

Figure 3.3: Line Graph — Enrollment Trends Over Time

In this example, the line graph illustrates how college enrollment changed year by year from 2015 to 2024. Each dot represents the total number of students enrolled that year, and the connecting lines show whether enrollment increased, decreased, or plateaued. This visual format is ideal for quickly assessing long-term trends, policy impacts, or institutional growth over time. It allows decision-makers and stakeholders to base strategies on historical patterns rather than guesses.

3.5 Cartesian Coordinate System

René Descartes (1596–1650) was a brilliant French philosopher, mathematician, and scientist whose intellectual legacy continues to shape both the humanities and the sciences. While he is perhaps most famously quoted for his statement *"Cogito, ergo sum"*—translated as **"I think, therefore I am"**—this phrase was not just a philosophical musing. It encapsulated his foundational belief in the primacy of reason and rational thought as the bedrock of knowledge. Descartes sought to establish a firm ground for all sciences, and in doing so, he developed methods that would revolutionize both philosophy and mathematics.

In mathematics, Descartes made his most enduring mark through the invention of the **Cartesian coordinate system**, which he introduced in his 1637 work, *La Géométrie.* This innovation laid the groundwork for analytic geometry—a field that blends algebra and geometry—and allowed equations to be visualized spatially on a plane. The Cartesian system uses two perpendicular number lines: the **x-axis**, which runs horizontally, and the **y-axis**, which runs vertically. Their intersection is the origin, denoted as $(0, 0)$, and the plane formed is divided into four quadrants.

What made Descartes' approach so revolutionary was the idea that geometric problems could be translated into algebraic equations, and algebraic relationships could be interpreted graphically. This fusion of geometry and algebra allowed for an entirely new way to understand mathematical relationships and paved the way for the development of calculus, physics, computer graphics, and countless applications in science and engineering.

Visualizing a Linear Relationship

The Cartesian coordinate system is not only useful for plotting shapes or solving geometric puzzles; it's a foundational tool in data analytics, economics, and business. One of its most powerful applications is in the visualization of **linear regression**, a method used to model the relationship between two quantitative variables.

Consider the linear equation:

$$y = 1.843x - 0.9943$$

In this model:

- x is the **independent variable** representing the amount of money spent on marketing, measured in thousands of dollars.
- y is the **dependent variable** representing the corresponding business revenue, also measured in thousands of dollars.

This equation tells us that for every additional \$1,000 spent on marketing, the company can expect an increase of approximately \$1,843 in revenue. The intercept, -0.9943, represents the theoretical revenue when no money is spent on marketing—though in practice, such values are often symbolic and must be interpreted within context.

Interpreting the Regression Line for Business Decisions

By plotting this regression line on a Cartesian plane, we gain more than just an equation—we gain a story. The visual representation shows a clear positive correlation between marketing expenditure and revenue. The upward slope demonstrates that as spending increases along the x-axis, the revenue values on the y-axis rise accordingly. This isn't just a mathematical relationship—it's a strategic insight.

For a business manager, this kind of analysis is invaluable. It provides quantitative justification for budget allocations. Instead of guessing or relying on anecdotal evidence, one can point to data-supported trends: more investment in marketing yields greater returns. This empowers decision-makers to act confidently, knowing their strategy is grounded in statistical reasoning.

Moreover, the visual clarity of the Cartesian plane aids in communication. Stakeholders, board members, and investors can easily interpret the trend when it is drawn as a line rising from left to right. This helps to build consensus around action plans and future projections. Visualizing regression models in this way is a cornerstone of data-driven decision-making in the modern economy.

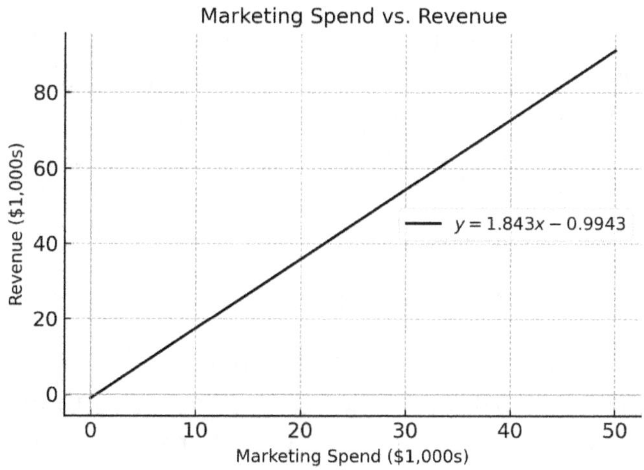

Figure 3.2 — Regression Line on a Cartesian Plane
This linear regression model visualizes how marketing investment (x-axis) influences business revenue (y-axis). The positive slope affirms that more marketing leads to increased income, making the case for strategic advertising stronger.

Understanding the Cartesian system is more than just knowing how to plot points. It's about seeing relationships, making predictions, and turning abstract numbers into concrete insights. Through Descartes' invention, we not only gained a mathematical tool—we inherited a visual language for interpreting the world.

3.6 Bonus: 3D Example

While two dimensions (x-axis and y-axis) are sufficient for many applications, the addition of a third dimension (z-axis) allows us to analyze more complex relationships—especially in business where multiple variables influence an outcome.

Understanding the 3D Cartesian Plane

In a three-dimensional Cartesian plane, we plot points using three coordinates: (x, y, z). These axes represent independent variables (x and z) and the dependent outcome (y). This format enables us to model how multiple factors together influence a result.

- The **x-axis** represents money spent on marketing.
- The **z-axis** represents customer engagement (e.g., website clicks, time on site, or social media interactions).
- The **y-axis** represents total revenue generated.

Examining Two Independent Variables

We use the following model to describe the relationship:

$$y = 2.1x + 1.3z + 5$$

- The coefficient **2.1** tells us that for every additional dollar spent on marketing, revenue increases by \$2.10—assuming customer engagement is held constant.

- The coefficient **1.3** means that for every unit increase in customer engagement, revenue increases by \$1.30—assuming marketing spend is held constant.
- The constant **5** represents baseline revenue when both marketing and engagement are zero.

Visualizing the Effect

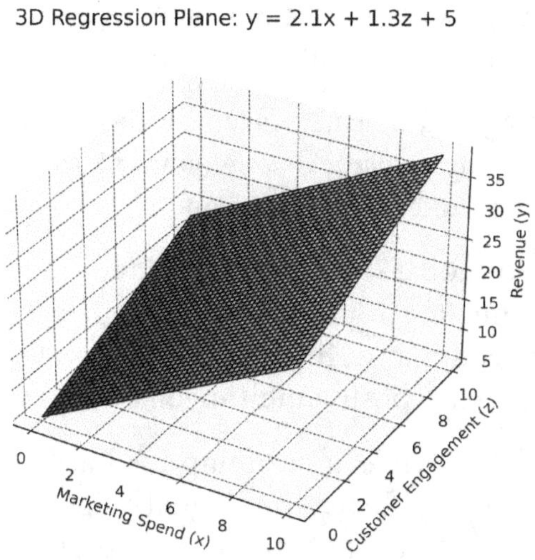

Figure 4.2 — Regression Surface in 3D
A 3D Cartesian plane illustrating the relationship between marketing (x), engagement (z), and revenue (y).

Managerial Interpretation

Visualizing this model in three dimensions helps us see how revenue is not only affected by marketing spend but also by how engaged the customers are. The regression surface slopes upward along both the x and z axes, showing a positive correlation between each variable and revenue.

As a business manager, this graph offers strategic insight:

- **Dual Investment Strategy:** You shouldn't only invest in marketing. You should also work to increase customer engagement (e.g., via better content, UX, loyalty programs).
- **Marginal Gains:** Even modest increases in either marketing or engagement result in measurable gains in revenue.
- **Balanced Growth:** The plane suggests that maximizing revenue is not about choosing between marketing and engagement—but about improving both simultaneously.

This model provides a roadmap for business decision-making: Invest intelligently in both external promotion and internal customer experience. The data tells a story of synergy—each dollar spent and each click earned together fuel your revenue engine.

3.7 How Visuals Lie

Not all charts are created honestly. A graph can be technically accurate but still deceptive. Here are some common tricks to watch out for:

Axis Manipulation

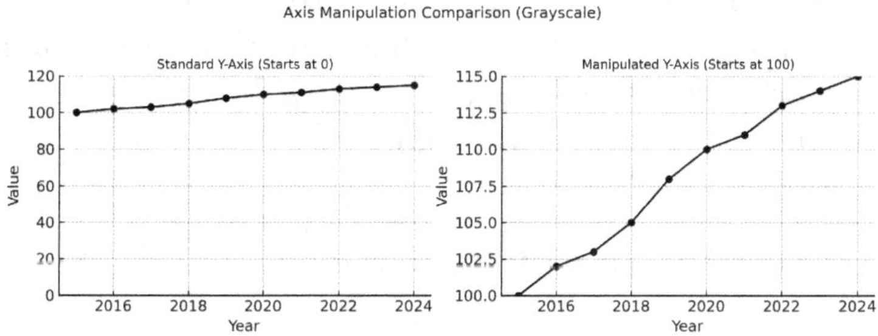

Figure — Axis Manipulation Illustrated
Two line graphs showing the same data, but the one on the right exaggerates the trend by starting the y-axis at 100 instead of 0.

Axis manipulation is one of the most common ways that visuals are used to mislead or exaggerate conclusions. By altering the scale of either axis—especially the y-axis—viewers may be tricked into seeing trends that aren't as significant as they appear.

In the graph on the left, the y-axis starts at 0. This is the most honest way to display data, as it provides the viewer with full context. The trend line gently rises over time, which correctly communicates a gradual increase in the data.

Now compare this with the manipulated graph on the right. By truncating the y-axis and starting it at 100, the same data appears to rise sharply, making the viewer think there's a dramatic increase—even though the change is relatively minor.

This kind of visual trickery is often seen in marketing, politics, and media presentations. Always check the axis scales when interpreting any chart. As a rule of thumb, if a chart seems too dramatic or too subtle to be true, inspect the axes before drawing any conclusions.

Inappropriate Chart Types

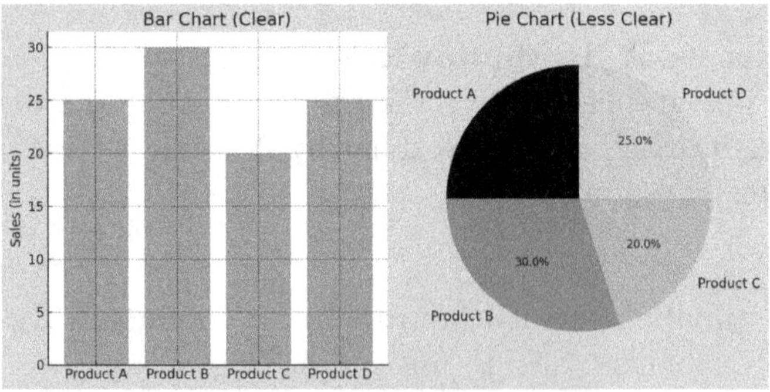

Figure — Comparing Chart Types: Bar vs. Pie A bar chart presents clear categorical comparisons, while the pie chart introduces perceptual distortion.

Not all chart types are equally effective at communicating data. Using a pie chart when a bar chart would be clearer can confuse viewers, especially when differences between values are small or when there are too many categories.

In the left chart above, the bar chart clearly displays the differences between products. Each bar is easy to compare by height, and values are readily understood. On the right, the pie chart uses angles and area to show proportions, but it becomes harder to detect the difference between similarly sized slices (e.g., Product A vs. Product D).

Additionally, adding unnecessary 3D effects can distort interpretation. A 3D pie chart or bar chart may cause certain segments or bars to appear larger or smaller than they are, due to visual perspective rather than actual value. For clear, accurate communication of data, always choose the simplest and most direct visual format.

Cherry-Picked Data

One of the most subtle and deceptive forms of data manipulation occurs through **cherry-picking**—selectively showing only a narrow window of the data that supports a specific narrative while omitting the broader context. This tactic can give a completely false impression of the actual trend or relationship in the dataset.

For example, imagine analyzing sales revenue over a 10-year period. If you only display the results from a 2-week dip during one year, you might incorrectly suggest that the company is failing—even though the long-term trend shows steady growth. This tactic is often used in media, advertising, and biased reports to push an agenda.

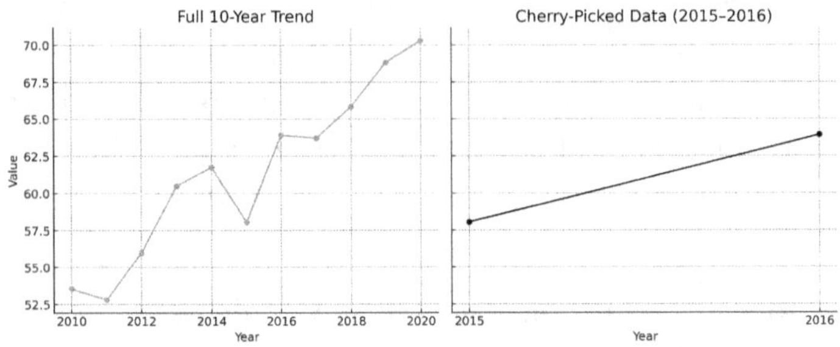

Figure 3.5 — The Danger of Cherry-Picked Data
Left: The full 10-year trend shows a steady upward pattern in revenue.
Right: A cherry-picked 2-week window (2015–2016) makes the trend appear flat or even negative.
This visual trick misleads by ignoring the bigger picture.

As a responsible analyst or business leader, it's essential to be aware of these tactics. Always ask:

- What is the full time frame of the data?
- Are there missing months, years, or observations?
- Does this slice accurately reflect the larger trend?

Cherry-picking breaks the trust between the presenter and the audience. Ethical data visualization requires honesty, context, and transparency. When used correctly, graphs can illuminate insights. When abused, they can mislead, distort, and manipulate perception.

Overuse

Too many colors, gradients, or patterns can distract from the actual data. Simpler is often clearer. Ask: Does this design enhance understanding or just add flair?

Conclusion

Charts and graphs are powerful storytelling tools. But like all tools, they can be used well—or poorly. The key is to know your data type, choose the right visual, and always aim for clarity and honesty.

In the next chapter, we'll build your skills in selecting, building, and evaluating visuals. You'll learn not just how to interpret charts—but how to create your own.

Check Your Understanding

In each of the following scenarios, identify the ethical problems with the graph or presentation. Consider how the person creating the graph may be attempting to deceive, mislead, or portray an untrue representation of the data. Write your answers in the space provided.

1. **Scenario 1 — Omitted Years in Sales Data:** A company presents a line graph showing "explosive growth" over three months. The graph starts in October and ends in December, with sales rising steadily. However, earlier that year, sales dropped drastically between January and

September.
Question: What ethical issues are present? How is the graph misleading?

2. **Scenario 2 — Y-Axis Manipulation in Crime Statistics:**
A city councilor presents a bar chart to show a dramatic drop in crime. The y-axis starts at 92 instead of 0, making the small decrease appear extreme.
Question: What effect does the truncated y-axis have on interpretation? Is this ethical?

3. **Scenario 3 — Inappropriate Chart Type in Budget Breakdown:**
A nonprofit uses a pie chart to show how their budget is spent. One slice labeled "operations" includes multiple unrelated expenses, and the colors used are hard to distinguish.
Question: What's problematic about the chart design? How could it be improved?

4. **Scenario 4 — Cherry-Picked Polling Results:**
A political ad displays a bar chart from a poll conducted during a holiday weekend with very low voter turnout. It excludes more comprehensive data from regular polling periods.
Question: What is wrong with this choice of

data? What does it fail to represent?

5. **Scenario 5 — Misleading Growth Projection:**
An investment firm shows a line graph predicting 200% portfolio growth in one year. The graph includes no error bars, confidence intervals, or mention of market risks.
Question: What key information is missing? Is this graph a fair representation of potential outcomes?

Discussion Prompt:
Why is it important to consider ethics when designing or interpreting data visualizations? How can being visually honest improve decision-making and trust?

Chapter Summary

In this chapter, we explored the power and pitfalls of data visualization. From bar charts to line graphs, we saw how visuals can clarify patterns, show trends, and make numbers meaningful. But we also learned how charts can lie—whether through manipulation of axes, inappropriate graph types, or cherry-picked data.

Key Takeaway: With great visual power comes great ethical responsibility. Always aim to inform, not mislead.

Chapter 4

How Data is Collected

Why Data Collection Matters

Before any analysis can begin, before any chart can be drawn or insight extracted, data must first be collected. This step is foundational—every conclusion we reach stands on the quality of our data collection. If the data is biased, incomplete, outdated, or gathered carelessly, then any conclusions drawn will be flawed or outright wrong. Even the most visually stunning graphs cannot rescue poorly collected data.

Imagine trying to cook a gourmet meal with spoiled milk or stale bread. That's what analyzing bad data is like. The method of data collection influences its reliability, validity, and generalizability. Well-collected data opens the door to accurate insights and responsible decision-making. Poorly collected data does the opposite—it leads to false assumptions, bad policies, and wasted resources.

Good data collection isn't just about numbers—it's about context. It's about ensuring the people, systems, or behaviors you're studying are properly understood. Think of data like ingredients in a recipe: even the best chef can't create a great dish from spoiled or unbalanced components. Similarly, even the most advanced machine learning models will fail if the data fed into them is faulty.

4.1 Data Collection

There are three primary approaches to gathering data in research, analytics, and business practice. Each method has distinct advantages and disadvantages. Understanding when and how to use each is crucial to producing reliable and actionable insights.

4.1.1 Surveys

Surveys gather data by asking people questions directly. They can be conducted face-to-face, over the phone, via paper forms, or—most commonly today—online through platforms like Google Forms, SurveyMonkey, or Qualtrics.

- **Strength:** Surveys are cost-effective and scalable. You can gather large amounts of data quickly from geographically dispersed respondents.

- **Weakness:** Responses depend on participant honesty and comprehension. Poorly worded questions or confusing formats can lead to inaccurate results.
- **Example:** A university sends an anonymous online survey to all students to assess satisfaction with campus facilities and instruction quality, using Likert scale responses from 1 (Strongly Disagree) to 5 (Strongly Agree).

Surveys are ideal for gathering subjective information—what people think, believe, or feel. However, they're prone to bias if the sample isn't diverse or if leading questions influence the outcome. Always test your survey instrument before widespread deployment.

4.1.2 Observational Studies

Observational studies involve watching subjects in their natural environment without interference. Researchers record behaviors, patterns, or outcomes without introducing experimental variables.

- **Strength:** Captures real-world data that reflects authentic behavior. Useful in sociological, medical, and retail studies.
- **Weakness:** Cannot determine causation. Observer presence may influence behavior (known as the Hawthorne effect).
- **Example:** A researcher in a retail store counts

how many customers stop to examine a promotional display and records their actions.

Observational studies are non-intrusive and provide a rich picture of what actually happens—versus what people say happens. Still, they require keen attention and often involve lengthy time commitments to ensure sufficient data is collected.

4.1.3 Experiments

Experiments are structured investigations where researchers manipulate one or more variables to determine their effect on another variable. Participants are typically split into at least two groups: a treatment group (which receives the intervention) and a control group (which does not).

- **Strength:** Provides strong evidence for causation and allows researchers to isolate variables and eliminate confounding factors through random assignment.
- **Weakness:** Can be expensive, time-consuming, or unethical (depending on the intervention). The artificial nature of experimental settings may not reflect real-world conditions, potentially limiting generalizability.
- **Example:** A pharmaceutical company tests a new drug by giving it to one group of participants while another receives a placebo. Both groups are monitored over several weeks for changes in symptoms and side effects.

Experiments are considered the gold standard for scientific research. Randomized controlled trials (RCTs), in particular, minimize bias and establish firm links between cause and effect by using randomization and blinding techniques. For example, double-blind studies—where neither the participant nor the researcher knows who receives the treatment—help prevent subconscious bias from influencing the results.

Beyond medicine, experiments are used in education, psychology, business, and even marketing. A/B testing on websites is a form of experimental design where different user experiences are tested to determine which results in more clicks or conversions. In agriculture, experiments might test how different fertilizers affect crop yield. In social psychology, researchers might study how group dynamics influence decision-making.

However, ethical constraints and logistical hurdles can limit the applicability of experiments. Some questions simply cannot be tested ethically—for instance, you cannot randomly assign children to poverty to study long-term effects. Similarly, controlled environments may strip away contextual nuance, making the results less applicable to real-life situations.

Moreover, experiments require careful planning, proper sample sizing, and robust statistical analysis to ensure results are meaningful and replicable. Missteps at any stage—such as failing to control confounding variables, not accounting for placebo effects, or interpreting correlation as causation—can invalidate the results.

Despite these challenges, experiments remain one of the

most powerful tools in the researcher's arsenal. They allow us to go beyond description and correlation to truly understand why something happens, making them indispensable for theory building, policy design, and evidence-based practice.

4.2 Ethics

Collecting data is not just a technical process—it is also a moral one. Ethical considerations are critical, particularly when dealing with sensitive subjects, vulnerable populations, or personally identifiable information.

Ethics in data collection is not merely a checklist to satisfy regulatory bodies—it is a living philosophy that must be woven into every stage of a research or analytical process. From deciding who to include in your study to how the data will be stored and analyzed, ethical decision-making can mean the difference between meaningful, responsible research and harmful, exploitative practice.

One of the foundational principles of ethical data collection is respect for persons. This includes respecting individual autonomy, obtaining informed consent, and giving participants the right to withdraw at any time. Researchers and data scientists must recognize that behind every data point is a human being with rights, dignity, and a story. This is especially important in studies involving minors, individuals with cognitive disabilities, or marginalized communities.

Confidentiality is another ethical cornerstone. Sensitive data—such as medical histories, financial records, or personal identities—must be protected through encryption, secure storage, and anonymization techniques. Transparency is equally crucial. Participants should be told what is being collected, how it will be used, who will have access, and for how long the data will be stored.

Moreover, ethics includes justice and equity in representation. This means striving to avoid data practices that reinforce existing inequalities, and being proactive about inclusion. It also involves avoiding practices that commodify or exploit user behavior, especially in digital platforms where consent may be buried in unread terms and conditions.

Ultimately, ethical data collection isn't just about protecting participants—it's about preserving the integrity of the research process and ensuring that insights drawn from data lead to positive, equitable, and socially responsible outcomes.

4.2.1 Sampling Bias

Sampling bias occurs when the sample collected doesn't accurately reflect the population of interest. This results in skewed findings and reduces the ability to generalize results. Bias can enter unintentionally through poor design, bad timing, limited access, or systemic exclusion of certain demographic groups.

For example, collecting survey responses only from

individuals who attend a specific type of event (such as a professional conference) may exclude lower-income or rural participants who couldn't attend. Similarly, using an app-based survey without considering that older populations or people with limited internet access may be left out can drastically shape the conclusions.

Example: An app developer sends a satisfaction survey exclusively to premium users—ignoring the experience of free-tier users. The results paint an unrealistically positive picture, as they exclude potentially critical feedback from a large segment of the user base.

Sampling bias isn't just a technical error—it's an ethical one. When decision-makers use biased data to guide products, policies, or public health strategies, the effects ripple outward, reinforcing inequalities and distorting reality.

To reduce sampling bias, researchers must:

- Define the target population precisely and inclusively.
- Use random or stratified sampling techniques where possible.
- Actively seek input from hard-to-reach or underrepresented groups.
- Pilot test instruments in diverse communities to flag potential bias.
- Transparently report sampling methods and known limitations.

Bias in sampling can lead to misinformed investments,

inequitable service delivery, or flawed predictions in algorithmic systems. Especially in the era of big data and artificial intelligence, ensuring a representative sample is a safeguard not only for accuracy, but for justice.

4.2.2 Ethical Concerns

Researchers must prioritize informed consent, data privacy, and participant welfare. This includes ensuring that every participant fully understands what data is being collected, why it is being collected, who will have access to it, and how it will be used or stored. Ethical data collection is built on the foundation of transparency, respect, and protection.

Data should never be collected through deception unless it has been clearly justified, reviewed, and approved by an independent ethics board or Institutional Review Board (IRB). Even when such exceptions are granted, the deception must be minimal, justified by the research goals, and followed by a debriefing. The goal is always to minimize harm and preserve individual dignity.

Example: Collecting browsing data without user consent may be legally permissible in some jurisdictions but remains ethically questionable. Even if no laws are broken, such practices can violate user expectations and erode public trust.

Moreover, ethical concerns extend beyond the act of collection to include the storage, analysis, sharing, and potential future use of the data. Are participants' identities adequately protected? Are there safeguards against re-identification in anonymized datasets? Are third-party collaborators held to the same ethical standards?

In today's global data environment, researchers must navigate complex legal frameworks like the EU's General Data Protection Regulation (GDPR) and the U.S. Health Insurance Portability and Accountability Act (HIPAA). These laws codify principles such as purpose limitation, data minimization, right to access, and right to be forgotten. But ethical responsibility doesn't end at legal compliance—it extends to moral duty and the trust participants place in those who collect and analyze their information.

Informed consent should not be treated as a formality. Researchers must ensure that consent is voluntary, specific, and informed—meaning participants understand the nature of the research, the risks involved, and their right to withdraw at any time without penalty. When working with vulnerable populations (such as children, individuals with cognitive impairments, or economically disadvantaged groups), additional protections must be put in place.

Ethical lapses in data collection can lead to lasting harm—not just to individuals, but to entire communities. They can also discredit scientific work, lead to legal consequences, and damage institutional reputations. Upholding ethical standards is therefore essential for both moral and practical reasons.

Ultimately, ethical data collection is about respect:

respecting the autonomy, privacy, and dignity of every individual whose information we seek to understand. When researchers act with integrity, they not only protect their subjects but also enhance the quality and legitimacy of their findings.

4.3 Underrepresentation

When certain groups are left out of the data, the results lack fairness, accuracy, and inclusivity. Underrepresentation can distort outcomes, reinforce systemic biases, and lead to products or policies that fail large segments of the population. This issue is especially pronounced in fields such as healthcare, education, public policy, and artificial intelligence systems.

Example: A voice recognition app trained primarily on male English speakers may misinterpret or fail to recognize the speech patterns of women, elderly individuals, children, non-native speakers, or those with regional dialects. As a result, these users experience reduced access, frustration, or even discrimination in automated systems.

The ripple effects of underrepresentation extend far beyond the immediate users. For instance, in healthcare, a dataset that lacks representation of minority populations may result in treatments or diagnostic tools that are less effective—or even harmful—for those communities. In education, if studies only include data from urban schools, rural or marginalized student populations may not benefit from resulting policies or innovations.

To counteract underrepresentation, researchers and data scientists must make deliberate efforts to ensure diversity in their samples. This includes designing studies that are culturally sensitive, linguistically inclusive, and geographically broad. Outreach efforts should extend to underserved communities, and researchers should collaborate with stakeholders to identify and address barriers to participation.

Moreover, data collection tools must be evaluated for inherent bias. Are surveys accessible to people with disabilities? Are language choices inclusive? Are incentives offered that may unintentionally exclude low-income participants? Thoughtful design and equitable implementation are essential.

Ultimately, building inclusive models and generating fair insights requires more than statistical rigor—it demands empathy, awareness, and an ethical commitment to justice. Without intentional inclusion, data becomes a mirror that reflects only a narrow slice of reality.

4.4 Groups

In statistics, it is essential to understand the distinction between a **population** and a **sample**. This distinction affects how data is collected, analyzed, and interpreted.

What Is a Population?

A **population** refers to the entire group of individuals, items, or observations that you want to study or draw conclusions about. Depending on the context, a population could include all citizens of a country, all products in a factory, or all website visits in a given year.

What Is a Sample?

A **sample** is a subset of the population. It includes only a portion of the entire group and is used when it is impractical or impossible to collect data from every member of the population.

Why the Distinction Matters

The distinction between a population and a sample is crucial because it determines:

- The statistical techniques used for analysis
- The type of inferences you can make
- The validity and reliability of your conclusions

Since samples are only a portion of the population, they can lead to sampling error. This means the characteristics of the sample may not perfectly represent those of the population. The smaller the sample, the greater the risk of this mismatch. Therefore, careful planning of sampling methods is necessary to reduce bias and increase representativeness.

Probability vs. Non-Probability Sampling

Sampling methods can be divided into two major categories: **probability sampling** and **non-probability sampling**.

Probability Sampling

In **probability sampling**, each member of the population has a known and non-zero chance of being selected. This allows for more accurate and generalizable results.

- **Simple Random Sampling:** Every member of the population has an equal chance of being selected. It is the gold standard for unbiased sampling but can be difficult to implement without a full list of the population.
- **Systematic Sampling:** Participants are selected at regular intervals from an ordered list. For example, every 10th name in a roster. This is simpler than random sampling but can introduce bias if there's an underlying pattern in the list.
- **Stratified Sampling:** The population is divided into distinct subgroups (strata) based on characteristics like age or income, and then random samples are drawn from each stratum. This ensures each subgroup is fairly represented.
- **Cluster Sampling:** The population is divided into clusters (e.g., neighborhoods or schools), and a random selection of clusters is chosen. All or

some individuals within selected clusters are then surveyed. This method is cost-effective for large populations spread over wide areas.

Non-Probability Sampling

In **non-probability sampling**, some members of the population may have no chance of being included in the sample. This type of sampling is easier and faster to conduct, but less reliable for generalizing findings.

- **Convenience Sampling:** Participants are selected based on ease of access, such as surveying people in a shopping mall. While inexpensive, this method is prone to bias.
- **Quota Sampling:** Researchers ensure the sample reflects certain characteristics of the population, but participants are selected non-randomly. For example, choosing 50 men and 50 women to mirror gender ratios.
- **Purposive (Judgmental) Sampling:** Participants are handpicked because they possess specific traits, knowledge, or experience relevant to the study.
- **Snowball Sampling:** Current participants recruit future participants. This method is often used for studying hidden or hard-to-reach populations, such as people with rare diseases or underground social groups.
- **Consecutive Sampling:** All eligible individuals are included over a defined time period or until a

target sample size is reached. This method reduces selection bias compared to convenience sampling.

- **Voluntary Response Sampling:** Individuals opt-in to participate, typically in response to an open call, survey, or media invitation. This can attract people with strong opinions and lead to skewed results.

Key Takeaways

- The choice between studying a population or a sample depends on resources, scope, and the goals of the study.
- Probability sampling yields more reliable, generalizable insights, but is often more complex and resource-intensive.
- Non-probability sampling is easier to implement but risks greater bias and reduced validity.
- When using samples to make inferences about populations, transparency in the sampling method and acknowledgment of limitations are essential for ethical and credible analysis.

Understanding the distinction between populations and samples—and selecting the appropriate sampling method—lays the foundation for sound statistical reasoning and evidence-based decision-making.

Probability Sampling Methods	
Method	**Description**
Simple Random Sampling	Every member of the population has an equal chance of being selected.
Systematic Sampling	Members are selected at regular intervals from a list or sequence.
Stratified Sampling	Population is divided into subgroups (strata), and random samples are drawn from each.
Cluster Sampling	Population is divided into clusters, and entire clusters are randomly selected.

Table 4.1: Probability Sampling Methods

Non-Probability Sampling Methods	
Method	**Description**
Convenience Sampling	Participants are selected based on ease of access and availability.
Quota Sampling	Aims to mirror population characteristics in fixed proportions.
Purposive Sampling	Participants are handpicked based on relevance or expertise.
Snowball Sampling	Participants refer others for inclusion in the study.
Consecutive Sampling	Includes all subjects meeting inclusion criteria over a time period.
Voluntary Response Sampling	People self-select to participate, often via open invitation.

Table 4.2: Non-Probability Sampling Methods

4.5 Looking Ahead

In this chapter, we laid the foundation for working with data by exploring how it is collected, the critical distinction between populations and samples, and the range of sampling methods used to extract meaningful insights. We examined both probability and non-probability sampling strategies, highlighting the strengths and limitations of each approach.

As we move forward, we'll build on this foundation by focusing on how to summarize and describe data using statistical tools. Understanding central tendency, variability, and distributional shape allows us to move from raw data to clear insights. By mastering these descriptive statistics, you'll gain the ability to interpret datasets with precision and prepare for more advanced analysis such as inference, prediction, and modeling.

Reliable data begins with rigorous collection and thoughtful sampling. From here, we deepen our toolkit to transform that data into knowledge—and eventually, into wisdom-driven decisions.

Chapter 5

Descriptive Statistics

Overview

Before we dive into complex statistical modeling, machine learning, or predictive analytics, we need to understand the basics—what the data tells us at a glance. Descriptive statistics are the first tools we use to make sense of a dataset. They are the statistical equivalent of reading a summary before diving into the full story, giving us the highlights without overwhelming details.

Descriptive statistics help us to summarize, describe, and understand the key features of a dataset, without making assumptions or predicting future trends. These statistics include tools that allow us to measure the center of the data, the spread of the values, and the general shape of the distribution.

They serve multiple purposes: providing a reality check for assumptions, offering insights for further investigation, and simplifying complex data so others can understand it. They also act as a diagnostic tool, helping us detect errors, outliers, and skewness early on in the analysis process.

Beyond that, descriptive statistics build the foundation for communication. Whether you're preparing a business report, a research paper, or a policy briefing, being able to communicate the story your data is telling is key. And often, the story begins with a handful of simple, elegant metrics.

Used correctly, descriptive statistics illuminate patterns, confirm expectations, and bring the data to life. They enable us to step back and say, "Here is what we know for sure before we move forward into deeper analysis."

In this chapter, we will examine the most common descriptive statistics—mean, median, mode, range, variance, and standard deviation—as well as understand when each one is appropriate to use depending on the type and shape of the data. We will do this without diving into complicated math formulas, instead focusing on interpretation, logic, and practical application.

By the end, you will have a strong understanding of how to interpret datasets, uncover meaningful insights, and communicate findings with confidence—all using the essential tools of descriptive statistics.

5.1 Summation

The concept of **summation** is fundamental in statistics and mathematics. It refers to the process of adding a series of numbers together, and it is commonly expressed using the Greek capital letter sigma ($\sum$), which stands for "sum."

What Does $\sum$ Mean?

The summation symbol $\sum$ tells you to add up a sequence of terms. It is usually written in the following format:

$$\sum_{i=1}^{n} x_i$$

This expression means: "Add up all the values of x from $i = 1$ to $i = n$."

- The $i = 1$ at the bottom tells you where to start.
- The n at the top tells you where to end.
- The x_i tells you the value to be added for each step.

Example 1: Simple Addition

Suppose we have the following numbers: 2, 4, 6, 8. These can be represented as $x_1 = 2$, $x_2 = 4$, $x_3 = 6$, and $x_4 = 8$.

Then:

$$\sum_{i=1}^{4} x_i = x_1 + x_2 + x_3 + x_4 = 2 + 4 + 6 + 8 = 20$$

Example 2: Indexed Variables

Let's say we define a variable $x_i = i^2$, and we want to find the sum of x_i for i from 1 to 5.

That is:

$$\sum_{i=1}^{5} i^2 = 1^2 + 2^2 + 3^2 + 4^2 + 5^2 = 1 + 4 + 9 + 16 + 25 = 55$$

Why Summation Matters

Summation is a building block of many statistical formulas, including the calculation of the **mean**, **variance**, and **standard deviation**. For example, the formula for the mean uses summation to add all the data points before dividing by the number of values:

$$\text{Mean} = \frac{1}{n} \sum_{i=1}^{n} x_i$$

Understanding how to read and use the $\sum$ symbol will

help you interpret formulas more confidently and apply them correctly in data analysis.

Tip for Business Readers

Even if you're not doing the math by hand, recognizing the $\sum$ symbol helps you understand how dashboards, reports, and software tools calculate things like totals, averages, or trends. If you're a manager, knowing this will give you greater insight into how your KPIs are built—and whether they make sense.

5.2 Mean, Median, Mode

5.2.1 Mean

The **mean**, commonly referred to as the *average*, is a statistical measure used to describe the central tendency of a dataset. It is calculated by summing all the numerical values in the dataset and then dividing by the number of values. The mean provides a single value that attempts to represent an entire set of values and is one of the most frequently used statistics in both academic research and practical business analysis.

Formula:

$$\text{Mean} = \frac{\sum x_i}{n}$$

Where x_i are the individual values in the dataset, and n is the total number of values.

Example:

Suppose a small business tracks the number of products sold each day over one work week: [20, 22, 19, 21, 100]. To find the mean:

$$\text{Mean} = \frac{20 + 22 + 19 + 21 + 100}{5} = \frac{182}{5} = 36.4$$

At first glance, this average may suggest that the business regularly sells over 36 items a day. However, the large value of 100 on one day (perhaps due to a big event or bulk order) skews the average, giving a distorted picture of the typical daily sales.

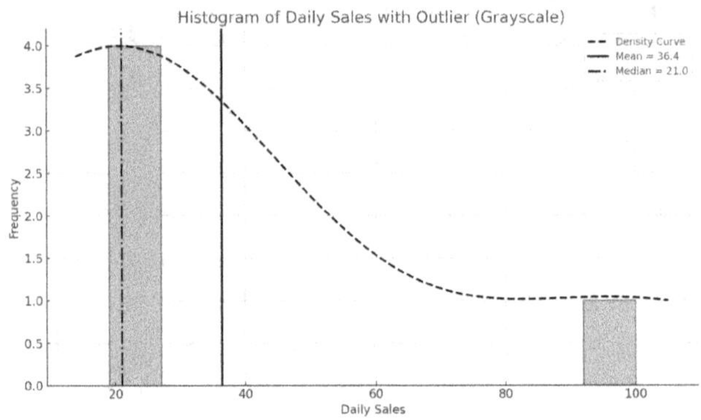

Figure 5.1: Distribution of Daily Product Sales With an Outlier

Figure 5.1 shows a histogram of daily product sales. Most days show sales between 19 and 22 units, but one day had a dramatic spike to 100 units. This outlier distorts the **mean**, pulling it to the right of the distribution's center.

In this example, the **mean** is approximately 36.4, while the **median** is only 21. The skew caused by the single high value shows how the mean can be misleading in a dataset with extreme values. The dashed vertical line marks the mean; the dash-dotted line shows the median.

5.2.2 Skewness

Skewness refers to the asymmetry of a probability distribution. If a distribution is symmetric, the mean and median are equal, and the skewness is zero. Positive skewness (right skew) indicates that the right tail is longer, and the mean is greater than the median. Negative skewness (left skew) occurs when the left tail is longer and the mean is less than the median.

- **Right (Positive) Skew:** Mean > Median
- **Left (Negative) Skew:** Mean < Median
- **No Skew:** Mean ≈ Median

In our histogram, the outlier (100) results in a *positively skewed* distribution, demonstrating how one extreme value can distort the perception of "average." The graph would be described as being skewed to the left.

5.2.3 Kurtosis

Kurtosis measures the "tailedness" or sharpness of the peak in a distribution. It indicates how much of the variance is due to infrequent extreme deviations (outliers), as opposed to frequent modest deviations.

- **Leptokurtic:** High kurtosis, with heavy tails and a sharp peak. Indicates a higher likelihood of extreme values.
- **Platykurtic:** Low kurtosis, with light tails and a flat peak. Indicates data are spread out evenly.
- **Mesokurtic:** Normal kurtosis, similar to a normal (bell-shaped) distribution.

In our example, the presence of an extreme outlier may increase the kurtosis, signaling a *leptokurtic* distribution. This highlights that unusual spikes in data—not just their asymmetry—can change how we interpret the entire dataset's shape.

Conclusion: Skewness and kurtosis help contextualize the mean and median by revealing hidden structure in the dataset. When used together, these metrics ensure analysts don't misread the central tendency or make flawed conclusions from atypical data points.

Usefulness

The mean is particularly useful when the data distribution is relatively symmetric and free of significant outliers. This is because extreme values can heavily influence the mean, pulling it in the direction of the skew. For instance, in a dataset of household incomes, a few billionaires could inflate the mean income in a way that doesn't accurately represent what most people earn.

Use it when: You're working with *continuous*

numerical data and want to describe the *typical or average value* of the dataset in a simple, interpretable way. It's commonly applied in fields such as economics (average salary), education (average test score), and marketing (average customer spend).

Understanding the mean helps in assessing general performance or identifying benchmarks. For example, a business owner calculating the mean monthly revenue can assess how well the business performs over time. If the current month's revenue falls far below the mean, it might indicate a problem that needs attention. Conversely, a revenue spike above the mean could suggest successful strategies worth repeating.

It is also useful in decision-making and forecasting. For example, if the mean customer satisfaction rating over the last twelve months is 4.2 out of 5, this can inform quality assurance strategies and identify patterns that influence customer perception.

The mean also plays an important role in comparative analysis. Suppose a school district wants to compare the performance of two schools. By calculating the mean test scores from each school, administrators can identify disparities, investigate root causes, and direct funding or intervention more effectively.

In business settings, tracking the mean values over time—such as average daily sales or average call wait times—can reveal trends and support data-driven decisions. Managers may set targets based on mean performance or use deviations from the mean to identify areas needing improvement.

However, it's crucial to interpret the mean in conjunction with other measures such as the median and standard deviation. This ensures a fuller understanding of the data's shape, variability, and potential skewness. Without this context, relying solely on the mean could lead to oversimplified or misleading conclusions.

For example, in real estate, the mean price of homes in an area can be significantly influenced by a few multimillion-dollar mansions, giving buyers or investors a distorted impression of affordability. In such cases, the median might provide a more reliable central tendency.

Ultimately, the mean is not just a number—it's a powerful summary that, when used appropriately, offers clarity and insight into complex datasets. It is both a starting point and a checkpoint, helping analysts, educators, policymakers, and business leaders interpret what the data is trying to say.

5.2.4 Median

The **median** is a measure of central tendency that represents the middle value in a dataset when all values are arranged in order from least to greatest. Unlike the mean, the median is unaffected by extremely high or low values, making it a more *robust* and reliable indicator of typical value in datasets that are skewed or contain outliers.

To find the median, one simply organizes the data in ascending order and identifies the middle number. If there is an odd number of values, the median is the center number. If there is an even number of values, the median is the average of the two center numbers.

Formula (conceptual): For a dataset with n values ordered from smallest to largest:

- If n is odd: Median = value at position $\frac{n+1}{2}$
- If n is even: Median = average of values at positions $\frac{n}{2}$ and $\frac{n}{2}+1$

Example: Consider the following dataset representing monthly rent prices in a neighborhood (in dollars):

$$800,\ 825,\ 850,\ 870,\ 900,\ 920,\ 950$$

Since there are 7 values (an odd number), the median is the fourth value when ordered:

$$\text{Median} = 870$$

Now consider a modified dataset that includes an outlier:

$$800,\ 825,\ 850,\ 870,\ 900,\ 920,\ 5000$$

Let's calculate the mean step-by-step in a wrapped format:

$$\text{Mean} = \frac{800 + 825 + 850 + 870 + 900 + 920 + 5000}{7}$$
$$= \frac{9165}{7} \approx 1309.29$$

$$\text{Median} = 870$$

As shown, the mean is significantly inflated by the presence of the outlier (5000), whereas the median remains at 870 and more accurately reflects the central tendency of the majority of the data. This highlights why the median is preferred when dealing with skewed data or extreme values.

Use it when: You have *skewed data*, such as income, property values, or hospital wait times, and you want to reduce the influence of extreme values. The median is especially helpful when working with data that is not normally distributed or contains major outliers.

In public policy and economics, median income is used more often than average income because it gives a clearer picture of what a "typical" person earns, regardless of extreme wealth at the top. In healthcare, median survival time is used in clinical trials to show how long patients typically live after treatment, avoiding distortion by a few long-living participants.

From a communications perspective, the median is often easier for audiences to understand and relate to. It aligns more closely with what most people experience, and it avoids the sensationalism that sometimes accompanies reporting based on mean values.

In education, a school district might report the median test score to represent student performance, since a few extremely low or high scores can warp the average. Similarly, real estate agents frequently quote median home prices, offering a clearer picture of market conditions and affordability.

Benefits of the Median:

- **Resilience to outliers:** It provides a better summary in the presence of skewed distributions.
- **Interpretability:** It aligns with the idea of a "typical" or "middle" value in everyday terms.
- **Fairer representation:** It avoids the distortion caused by extreme high or low values.

The median is not without limitations. It does not account for the size of gaps between values and cannot be used in further calculations as easily as the mean. But when the goal is fairness, robustness, and minimizing distortion, the median often proves more effective than the mean.

Ultimately, the median allows analysts to better understand distributions and to communicate insights in a way that acknowledges real-world variability. It complements the mean and should always be considered alongside it in any complete statistical analysis.

5.2.5 Mode

The **mode** is the value that appears most frequently in a dataset. It is a measure of central tendency that reflects the most common or popular item within a set of observations. Unlike the mean and median, which rely on numerical ordering, the mode simply identifies which value occurs with the greatest frequency, making it particularly valuable for categorical and nominal data.

A dataset may have:

- **No mode:** All values occur with the same frequency.
- **Unimodal:** One value appears most often.
- **Bimodal:** Two values appear with the same highest frequency.
- **Multimodal:** More than two values occur with equal highest frequency.

Use it when: You are analyzing *categorical or discrete data* and want to determine the most frequently occurring category, label, or value. It is useful in identifying dominant traits, behaviors, or preferences in a population.

Examples:

- In a survey about favorite ice cream flavors, if 45 people choose vanilla, 30 choose chocolate, and 25 choose strawberry, then vanilla is the mode.
- In a dataset of shoe sizes: the mode is 10.

The mode is especially informative when analyzing responses that cannot be averaged meaningfully. For instance, in market research, knowing the most popular brand, color, or feature preference can guide inventory and advertising strategies. In healthcare, the mode might indicate the most common symptom experienced by patients with a specific condition.

Advantages of the Mode:

- **Applicable to all data types:** Works with nominal, ordinal, and numerical data.
- **No calculation required:** Easily understood and identified from raw data.
- **Resilient to outliers:** Not affected by extreme values in the dataset.

Limitations of the Mode:

- May not exist (no mode) or may be unclear (multimodal).
- Provides limited insight into the spread or variability of the data.
- Less useful for further statistical computations.

In education, teachers might use the mode to determine the most commonly selected incorrect answer on a test, identifying areas where students need more instruction. In fashion retail, the mode of clothing sizes sold can help optimize stock levels.

Ultimately, while the mode is often overlooked in favor of the mean and median, it is an essential descriptive statistic. It gives voice to the majority and reveals the most common traits or outcomes in a dataset—insights that are often the most actionable in fields ranging from consumer behavior to epidemiology.

When used in combination with the mean and median, the mode enhances the overall understanding of data distribution, skewness, and central tendency, providing a multi-dimensional view of what the data truly represents.

5.3 Percentiles

Percentiles are measures that divide a dataset into 100 equal parts. Each percentile represents a relative standing of a value within a data set. For instance, the 20th percentile (denoted P_{20}) is the value below which 20% of the data points fall.

To compute the kth percentile (P_k), the formula used is:

$$\text{Position} = \frac{k(n+1)}{100}$$

where n is the number of observations in the dataset. This formula gives the position of the percentile value when the data are arranged in ascending order. If the position is not a whole number, interpolation is used between the two nearest ranks.

Example: Consider the dataset: 3, 7, 8, 12, 13, 14, 18, 21, 23, 27. To find P_{30}:

$$\text{Position} = \frac{30(10+1)}{100} = 3.3$$

So P_{30} lies between the 3rd and 4th values, which are 8 and 12. Interpolating:

$$P_{30} = 8 + 0.3(12-8) = 8 + 1.2 = 9.2$$

5.4 Quartiles

Quartiles divide a data set into four equal parts and are specific percentiles:

- Q_1 (first quartile) = 25th percentile
- Q_2 (second quartile) = 50th percentile (also the median)
- Q_3 (third quartile) = 75th percentile

These values help summarize the spread and center of a dataset.

Example: In the dataset from earlier:

- $Q_1 = P_{25} \approx 7.5$
- $Q_2 = \text{Median} = \frac{13+14}{2} = 13.5$
- $Q_3 = P_{75} \approx 21$

5.5 Deciles

Deciles divide data into ten equal parts. Each decile represents a 10% increment.

The kth decile (D_k) corresponds to the $k \cdot 10$th percentile. For instance, $D_1 = P_{10}$, $D_5 = P_{50}$, and $D_9 = P_{90}$.

Example: Using the same dataset:

$$D_4 = P_{40} = \text{Position} = \frac{40(11)}{100} = 4.4$$

Between 4th and 5th values (12 and 13):

$$P_{40} = 12 + 0.4(13 - 12) = 12.4$$

5.6 Range, Minimum, and Maximum

The **range** is a basic measure of variability, calculated as:

$$\text{Range} = \text{Maximum value} - \text{Minimum value}$$

Minimum: The smallest data value.
Maximum: The largest data value.

Example: For the dataset 3 to 27:

$$\text{Range} = 27 - 3 = 24$$

Range is simple to calculate and helps understand the extent of the data spread.

5.7 Interquartile Range (IQR)

The interquartile range (IQR) is a robust measure of variability, focusing on the middle 50% of the data:

$$\text{IQR} = Q_3 - Q_1$$

It is often used to detect outliers. An outlier is a value that is significantly different from others. IQR-based outlier criteria:

- Lower bound = $Q_1 - 1.5 \times \text{IQR}$
- Upper bound = $Q_3 + 1.5 \times \text{IQR}$

Example: If $Q_1 = 7.5$, $Q_3 = 21$, then:

$$\text{IQR} = 21 - 7.5 = 13.5$$

$$\text{Lower bound} = Q_1 - 1.5 \cdot \text{IQR} = 7.5 - 1.5 \cdot 13.5 = -12.75$$
$$\text{Upper bound} = Q_3 + 1.5 \cdot \text{IQR} = 21 + 1.5 \cdot 13.5 = 41.25$$

Any values outside $[-12.75, 41.25]$ are outliers.

5.8 5-Number Summary

The **five-number summary** consists of:

- Minimum
- First Quartile (Q_1)
- Median (Q_2)
- Third Quartile (Q_3)
- Maximum

Steps:

1. Sort the data in ascending order.

2. Identify the minimum and maximum values.
3. Find the median (Q_2).
4. Find Q_1 (median of the lower half, excluding Q_2 if n is odd).
5. Find Q_3 (median of the upper half).

5.9 Box Plot

A **box plot** (or box-and-whisker plot) is a graphical display of the five-number summary.

Steps:

1. Draw a number line covering the data range.
2. Draw a box from Q_1 to Q_3.
3. Draw a vertical line at the median inside the box.
4. Extend "whiskers" from Q_1 to the minimum and from Q_3 to the maximum.

Outliers may be plotted as individual points beyond the whiskers.

5.10 Whisker Plot

The **whisker plot** is synonymous with a box plot, but emphasizes the "whiskers" and potential outliers.

Additional Features:

- Any data point beyond $1.5 \times \text{IQR}$ from Q_1 or Q_3

is plotted as a separate dot.

- Whiskers extend to the last data point within the $1.5 \times$ IQR range.

This visualization helps detect skewness and extreme values in the dataset.

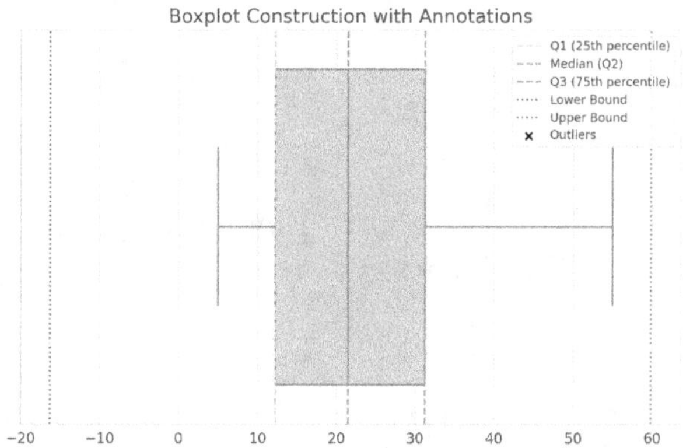

Figure 5.2: Boxplot Construction: Showing Q1, Median (Q2), Q3, Lower/Upper Bounds, and Outliers.

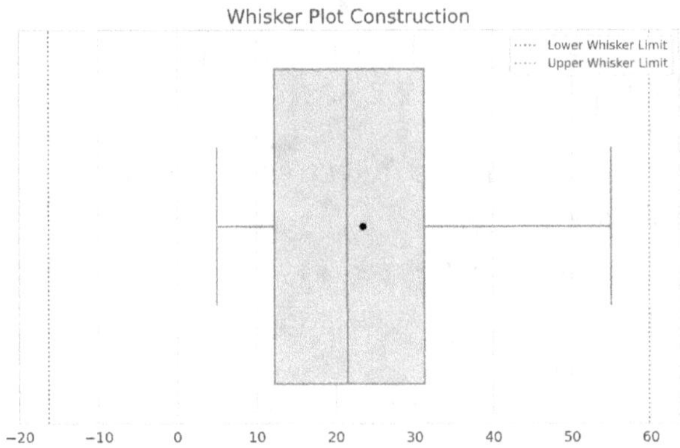

Figure 5.3: Whisker Plot: Emphasizing the whisker limits with outlier identification.

5.11 Centrality

Understanding the **center** of a dataset is essential because it gives us a reference point—a summary that reflects what is typical or expected in the data. Whether you're analyzing income levels, test scores, product reviews, or health outcomes, the center tells you what lies at the heart of your data.

But not all datasets behave the same way. Some are perfectly symmetrical; others are skewed or contain outliers. That's why we don't rely on a single method to measure centrality. Instead, we use three core metrics—**mean**, **median**, and **mode**—each offering a unique perspective.

- **Mean:** The arithmetic average, useful when data

is symmetric and free from extreme values. It is sensitive to outliers and skewed distributions.

- **Median:** The middle value in an ordered list. The median is more robust to outliers and skewed data, making it ideal when you want a typical value that isn't influenced by extremes.
- **Mode:** The most frequently occurring value. It's especially helpful with categorical data or when identifying the most common response or behavior in a dataset.

Each measure offers insights depending on the context:

- If you're analyzing salaries in a company with a few extremely high earners, the **median** gives a more realistic picture of typical employee earnings than the mean.
- In understanding test scores of a normally distributed classroom, the **mean** might be most informative.
- When analyzing shoe sizes or favorite ice cream flavors, the **mode** reveals the most popular option.

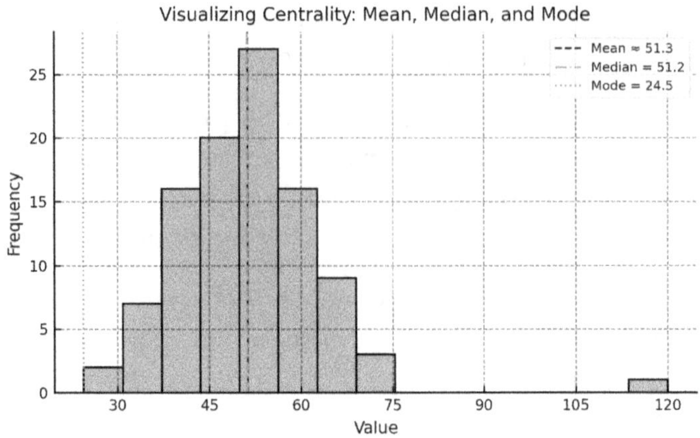

Figure 5.4: Visual comparison of mean, median, and mode across different distributions.

Choosing the right measure of centrality is not just about math—it's about accurately representing reality. It ensures that summaries are meaningful, decisions are based on fair representations, and communication is clear and intuitive. In essence, centrality helps us transform data from a list of numbers into an understandable story about what is usual, expected, or most common.

5.12 Parameters vs. Statistics

In statistics, it's crucial to distinguish between a **parameter** and a **statistic**.

A parameter is a numerical value that summarizes a characteristic of an entire population. Because parameters describe populations, they are often unknown and must be estimated. Greek letters are used to denote parameters.

A statistic is a numerical value that summarizes a characteristic of a sample. Since we usually can't measure every individual in a population, we rely on statistics to make educated guesses about the underlying parameters. Latin letters are used to denote statistics.

Example: Mean

Suppose we want to know the average (mean) height of all adult men in a country. The true mean of the entire population is a parameter, denoted by the Greek letter μ. However, if we take a sample of 1000 men and calculate the mean of their heights, this is a statistic, denoted by $\bar{x}$.

Formulas

Population Mean (Parameter):

$$\mu = \frac{1}{N}\sum_{i=1}^{N} x_i$$

Where:

- μ is the population mean,
- N is the total number of individuals in the population,

- x_i represents each individual value in the population.

Sample Mean (Statistic):

$$\bar{x} = \frac{1}{n}\sum_{i=1}^{n} x_i$$

Where:

- $\bar{x}$ is the sample mean,
- n is the number of individuals in the sample,
- x_i represents each individual value in the sample.

Why It Matters

This distinction is important because:

- Parameters are typically fixed but unknown.
- Statistics vary from sample to sample.
- We use statistics to make inferences about parameters.
- Many statistical formulas adjust based on whether we're analyzing a sample or a full population (e.g., variance uses $n - 1$ instead of n for samples).

Being clear about whether you are working with a population or a sample affects which symbols, formulas, and interpretations are correct in your analysis.

5.13 Spread

5.13.1 Variance

The **variance** is a fundamental statistical concept that quantifies how much the values in a dataset deviate from the mean. It measures the average squared deviation of each data point from the mean, giving us insight into the *spread* or *dispersion* of the data.

Why is variance important?

Variance tells us how consistent or inconsistent the values in a dataset are. A low variance means that the values are closely grouped around the mean—indicating uniformity or predictability. A high variance suggests that the data is more spread out, which can signal volatility, risk, or the presence of outliers.

Formula:

For a sample with values $x_1, x_2, \ldots, x_n$ and mean $\bar{x}$, the variance is calculated as:

$$s^2 = \frac{1}{n-1} \sum_{i=1}^{n} (x_i - \bar{x})^2$$

This is the average of the squared differences between each value and the mean.

Example:

Suppose we are looking at the number of hours five

students studied for a test:

$$4,\ 4,\ 5,\ 6,\ 11$$

First, calculate the mean:

$$\bar{x} = \frac{4+4+5+6+11}{5} = \frac{30}{5} = 6$$

Then calculate the squared deviations:

$$(4-6)^2 = 4, \quad (4-6)^2 = 4, \quad (5-6)^2 = 1, \quad (6-6)^2 = 0, \quad (11-6)^2 = 25$$

Sum of squared deviations:

$$4+4+1+0+25 = 34$$

Now divide by $n-1$:

$$s^2 = \frac{34}{4} = 8.5$$

So the variance is 8.5.

Analogy – The Campfire Circle:

Imagine five friends sitting around a campfire. If they all sit evenly spaced around the fire, the group has low variance—they're equally close to the center. But if one person sits far off in the woods, the group's arrangement has high variance. Even though the average distance to the fire might still seem reasonable, the spread of distances tells a very different story. Variance gives us that picture of how "tightly grouped" or "scattered" the data really is.

Units of Variance:

While variance is useful, it has a limitation: its units are the square of the original data units. If your dataset is in meters, the variance is in square meters (m^2); if your data is in dollars, variance is in dollars squared (USD^2). This squared unit makes it hard to directly compare the variance to the mean or to other datasets with different scales.

This leads us to a natural next step: the **standard deviation**.

Why go beyond variance?

To address the challenge of squared units, we often take the square root of the variance. This gives us the standard deviation—a measure that is in the same unit as the original data and is more interpretable. The standard deviation maintains the insight variance gives us, but expresses it in a scale we can better understand and communicate.

In summary:

- Variance tells us how widely spread our data is.
- High variance suggests inconsistency or potential outliers.
- Low variance indicates more uniform, consistent data.

- Because variance uses squared units, we often prefer standard deviation for interpretation.

Understanding variance is crucial for identifying how trustworthy our measures of central tendency (like the mean) really are. Without it, we might be misled by a deceptively "normal" average that masks major variation within the data. First, before we move on, we will talk about Population Variance and Sample Variance. Then we will proceed to work with Standard Deviation.

Population vs Sample

Variance is a fundamental concept in statistics that measures the degree to which individual values in a dataset differ from the mean. It captures the spread or dispersion of the data.

When calculating variance, it's important to distinguish whether you're working with an entire population or just a sample. This distinction affects how the formula is structured.

5.13.2 Population Variance

Population variance (σ^2) is used when you have data for every member of the population.

$$\sigma^2 = \frac{1}{N}\sum_{i=1}^{N}(x_i - \mu)^2$$

Where:

- σ^2 is the population variance,
- N is the total number of data points in the population,
- x_i represents each value,
- μ is the population mean.

5.13.3 Sample Variance

Sample variance (s^2) is used when your data is a subset (sample) of the full population. Because a sample may not capture all the variation in the population, we divide by $n - 1$ instead of n to correct for bias. This is known as *Bessel's correction.*

$$s^2 = \frac{1}{n-1}\sum_{i=1}^{n}(x_i - \bar{x})^2$$

Where:

- s^2 is the sample variance,
- n is the number of data points in the sample,
- x_i represents each value in the sample,
- $\bar{x}$ is the sample mean.

Key Distinction

- Use σ^2 when you know data for the entire population.

- Use s^2 when working with a sample, and adjust with $n-1$ to ensure an unbiased estimate.

Understanding whether you are computing a population or sample variance ensures accurate results and proper interpretation of data variability.

5.13.4 Standard Deviation

The **standard deviation** is a fundamental measure of dispersion that tells us how spread out the values in a dataset are from the mean. While measures like the mean and median help us understand the center of the data, standard deviation provides insight into the data's variability—how much the values deviate from that center.

In technical terms, the standard deviation is the square root of the variance. It takes the average of the squared differences between each data point and the mean, and then returns the square root of that average. The result is a value in the same units as the original data, making it easier to interpret than the variance itself.

Formula: Let $x_1, x_2, \ldots, x_n$ be a dataset of n values with mean $\bar{x}$. Then the standard deviation s (for a sample) is:

$$s = \sqrt{\frac{1}{n-1}\sum_{i=1}^{n}(x_i - \bar{x})^2}$$

Interpretation: - A *small* standard deviation means the values are close to the mean—indicating consistency or stability. - A *large* standard deviation means the values are more spread out—indicating greater variability or uncertainty.

Example – The River Analogy:

Imagine you are told that the average depth of a river is 3 feet. You might conclude it is safe to walk across. After all, 3 feet of water isn't typically dangerous for an adult. But this average is only part of the story.

Suppose the depths measured at several points in the river are:

$$0.1\text{ ft},\ 0.5\text{ ft},\ 1.0\text{ ft},\ 2.8\text{ ft},\ 3.2\text{ ft},\ 4.5\text{ ft},\ 9.0\text{ ft}$$

While the mean (average) might still be around 3 feet, the spread of values reveals a deeper problem. There's a location where the water is just 0.1 feet deep, and another where it plunges to 9 feet—well over most people's heads.

This is where the standard deviation becomes crucial. It highlights how inconsistent or dangerous the conditions may be, despite the reassuring average. In this case, a high standard deviation would alert us that the depth varies greatly—and that the river is unsafe to cross without further investigation.

Use it when: You want to understand how consistent the data is, how much risk or uncertainty is involved, or how much individual data points deviate from the mean. Standard deviation is widely used in fields like:

- *Finance:* to assess volatility of investments.

- *Quality control:* to ensure manufacturing consistency.
- *Healthcare:* to measure variation in clinical results.
- *Education:* to see how spread out test scores are.

Why It Matters: Averages can be deceptive. Two datasets can have the same mean but vastly different standard deviations. For example:

- Dataset A: 48, 49, 50, 51, 52 (Mean = 50, SD = low)
- Dataset B: 10, 30, 50, 70, 90 (Mean = 50, SD = high)

Without knowing the spread, we might falsely assume both datasets are similarly distributed around the center. But only by looking at the standard deviation can we grasp the full story.

Key Insight:

> *The mean tells you where the center is, but the standard deviation tells you how safe it is to rely on that center.*

Thus, standard deviation is not just a mathematical formula—it's a lens through which we understand uncertainty, reliability, and real-world decision-making.

5.14 Population SD, Sample SD

Standard deviation is one of the most widely used measures of variability in statistics. It quantifies the average amount by which values in a dataset deviate from the mean. It is simply the square root of the variance, returning the measure to the original unit of the data, which makes it easier to interpret.

Population Standard Deviation

Population standard deviation (σ) is used when you have data for the entire population.

$$\sigma = \sqrt{\frac{1}{N}\sum_{i=1}^{N}(x_i - \mu)^2}$$

Where:

- σ is the population standard deviation,
- N is the number of values in the population,
- x_i is each individual data point,
- μ is the population mean.

Sample Standard Deviation

Sample standard deviation (s) is used when your dataset represents a sample drawn from a larger population. To correct for the underestimation of variability, we divide by $n - 1$ instead of n, as done in sample variance.

$$s = \sqrt{\frac{1}{n-1}\sum_{i=1}^{n}(x_i - \bar{x})^2}$$

Where:

- s is the sample standard deviation,
- n is the number of values in the sample,
- x_i is each sample data point,
- $\bar{x}$ is the sample mean.

5.14.1 Interpretation

Standard deviation helps us understand the consistency or dispersion of the data:

- A **small standard deviation** means most values are close to the mean — low variability.
- A **large standard deviation** indicates the data is more spread out — high variability.

Visual example: In a dataset of exam scores with a standard deviation of 2, most students scored near the average. But if the standard deviation is 20, scores were widely spread — some very low and others very high.

Why It Matters

Understanding standard deviation allows for better comparison of datasets and helps in evaluating risk, consistency, and reliability. In research, business, healthcare, and engineering, standard deviation plays a vital role in interpreting how "typical" the mean truly is in representing the data.

5.15 Visualizing Variance

Understanding **variance** and **standard deviation** is not merely academic—it is crucial for interpreting how well a model fits data. Suppose we have a linear model:

$$y = 5x + 4$$

We might think this is an excellent model for predicting outcomes. However, the reliability of this model depends on how closely the actual data points follow this line.

Two Graphs: Low vs. High Variance

To visualize this, we plot two graphs below. Both use the same underlying model $y = 5x + 4$, but the data differs.

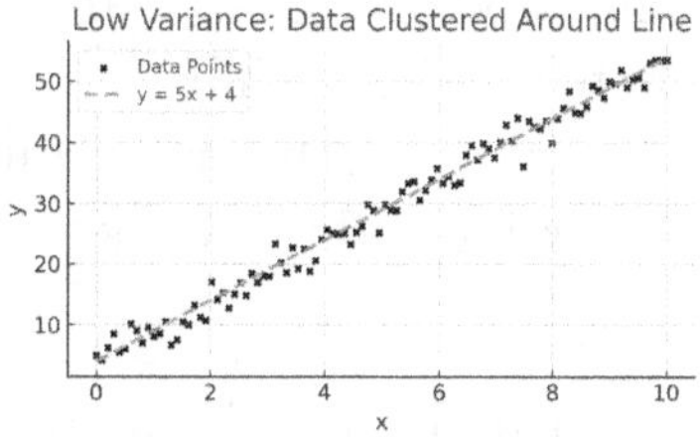

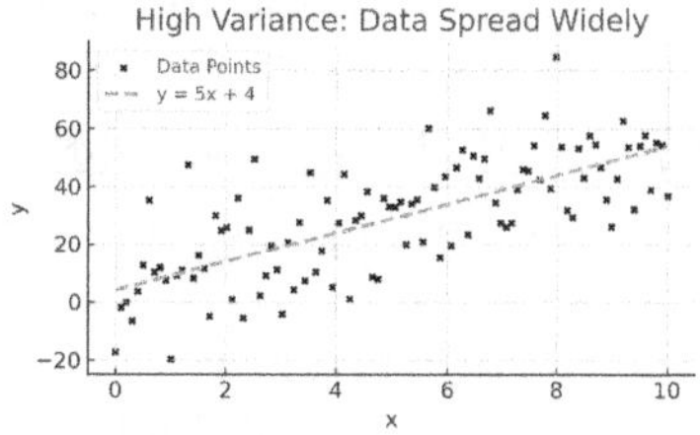

Left: Low Variance — The data points cluster

tightly around the regression line. This indicates:

- Low **residuals** (differences between actual and predicted values),
- Small **standard deviation**, and
- A high-quality model with good predictive power.

Right: High Variance — The data points are scattered widely around the regression line. This suggests:

- Larger residuals,
- High standard deviation, and
- A model that may not explain the data well.

Interpretation

These graphs demonstrate why summary statistics matter. A regression line alone doesn't reveal whether it's a good fit—you must examine the spread of the data. Standard deviation tells you how much the actual outcomes deviate from the model's predictions, while variance tells you how spread out the data is overall.

When data points lie close to the model line, we can trust the predictions more. When they're widely scattered, the model might still give the right trend but fail to give accurate predictions for individual data points.

Always graph your model against the data and calculate standard deviation and variance. Numbers and visuals together tell a complete story.

5.16 Looking Ahead

Now that we understand how to summarize data, the next step is to understand how it was sampled. In the following chapters, we'll explore how to draw conclusions from samples using inferential statistics.

5.17 Practice Exercises

Strengthen your understanding of central tendency and variability by solving the following exercises.

Exercise 1: Mean, Median, and Mode

Given the following dataset representing the number of books read by 10 students in a month:

$$4,\ 7,\ 6,\ 9,\ 7,\ 5,\ 8,\ 7,\ 6,\ 10$$

1. Calculate the **mean** (average) number of books read.
2. Find the **median** value of the dataset.
3. Determine the **mode**, if any exists.

Exercise 2: Sample Mean, Sample Variance, and Sample Standard Deviation

Consider the following sample dataset representing the weekly spending (in dollars) of 6 individuals on groceries:

$$55, \ 60, \ 52, \ 63, \ 58, \ 72$$

1. Compute the **sample mean** ($\bar{x}$).
2. Calculate the **sample variance** (s^2).
3. Compute the **sample standard deviation** (s).

Note: Use the formulas provided in earlier sections to guide your calculations.

Additional Exercises

1. Define the difference between a population and a sample. Give one example of each.
2. State whether each of the following is a parameter or a statistic:
   (a) The average age of all students at a university.
   (b) The average age of students in a statistics class of 35 students.
3. Given the data set: 3, 7, 7, 2, 10, 4, 4, 5, 9, 6, compute the mean, median, and mode.

4. Consider the data: 10, 12, 11, 13, 95. Explain how the presence of the value 95 affects the mean, median, and mode.

5. For the following set of data: 5, 8, 10, 12, 15, calculate the range, variance, and standard deviation.

6. A data set has a mean of 30 and a standard deviation of 5. Compute the coefficient of variation.

7. A grouped frequency distribution is provided:

Class Interval	Frequency
10–19	2
20–29	6
30–39	8
40–49	4

   Estimate the mean of the data set.

8. A data set has a long right tail. Is the distribution positively skewed, negatively skewed, or symmetric? Explain your answer.

9. Provide the five-number summary and sketch a boxplot for the data: 2, 4, 5, 7, 8, 12, 13, 16, 20, 21.

10. Explain the difference between computing variance for a sample and a population. Then compute both for the dataset: 6, 8, 10, 14.

Chapter 6

Probability Concepts

6.1 Understanding Probability

Probability is a numerical measure that describes the likelihood of an event occurring, and it always lies between 0 and 1, inclusive. A probability of 0 means the event is impossible—it cannot happen under any circumstances. A probability of 1 indicates certainty—the event will definitely occur. Every event you can imagine falls somewhere along this spectrum, from highly unlikely to inevitable.

What is an Event?

In probability theory, an **event** is a specific outcome or a group of outcomes from a random process. For instance, tossing a coin and getting heads is an event.

Rolling a die and getting a number less than 4 is another event. Events can be:

- **Simple**: Consisting of only one outcome (e.g., rolling a 6 on a die).
- **Compound**: Consisting of more than one outcome (e.g., rolling an even number).

Sample Space

The **sample space** is the set of all possible outcomes of an experiment. For example:

- Tossing a coin: Sample space = {Heads, Tails}
- Rolling a six-sided die: Sample space = {1, 2, 3, 4, 5, 6}

Each outcome in the sample space has a probability assigned to it, and the sum of all probabilities in the sample space equals 1.

Basic Probability Formula

If all outcomes in the sample space are equally likely, then the probability of an event is calculated as:

$$P(\text{Event}) = \frac{\text{Number of favorable outcomes}}{\text{Total number of possible outcomes}}$$

Example: What is the probability of drawing a red card from a standard deck of 52 cards?

$$P(\text{Red}) = \frac{26}{52} = 0.5$$

Probability Scale

The probability scale helps visualize how likely an event is:

- 0 = Impossible (e.g., rolling a 7 on a six-sided die)
- 0.5 = Equally likely to occur or not occur (e.g., flipping heads on a fair coin)
- 1 = Certain (e.g., drawing a card from a standard deck that is either red or black)

Relative Frequency as Probability

Probability can also be interpreted empirically as a **relative frequency**:

$$P(\text{Event}) \approx \frac{\text{Number of times event occurs}}{\text{Total number of trials}}$$

This is useful in experimental or real-world settings, especially when theoretical probabilities are difficult to determine.

Example: Suppose you observed 200 cars passing a toll booth, and 80 were trucks. Then the empirical probability that the next car is a truck is:

$$P(\text{Truck}) \approx \frac{80}{200} = 0.4$$

Why Probability Matters

Understanding probability is essential because it:

- Allows us to quantify uncertainty and make informed decisions.
- Forms the foundation for statistics, risk analysis, data science, and AI.
- Helps model real-world phenomena, from weather forecasts to quality control.

Events and Their Complements

The **complement** of an event is the set of outcomes in the sample space that are not in the event. If A is an event, its complement is denoted A' or A^c. The probabilities of an event and its complement always add up to 1:

$$P(A) + P(A') = 1$$

Mutually Exclusive Events

Two events are **mutually exclusive** if they cannot occur at the same time. For example, when rolling a single die, getting a 3 and a 5 are mutually exclusive events.

Addition Rule for Mutually Exclusive Events:

$$P(A \text{ or } B) = P(A) + P(B)$$

Non-Mutually Exclusive Events

When events can occur at the same time, they are not mutually exclusive. In such cases, you must subtract the probability of both events occurring together to avoid double-counting:

$$P(A \text{ or } B) = P(A) + P(B) - P(A \text{ and } B)$$

Independent vs. Dependent Events

- **Independent events:** The outcome of one event does not affect the outcome of another. (e.g., flipping a coin and rolling a die)
- **Dependent events:** The outcome of one event does influence the outcome of another. (e.g., drawing two cards from a deck without replacement)

Multiplication Rule

- For independent events:
  $P(A \text{ and } B) = P(A) \cdot P(B)$
- For dependent events:
  $P(A \text{ and } B) = P(A) \cdot P(B|A)$, where $P(B|A)$ is

the probability of B given A

Conclusion

Probability is more than just a mathematical concept—it's a lens through which we interpret the world. From simple coin tosses to complex medical diagnoses, understanding probability helps us navigate uncertainty, reduce risk, and approach life with clearer expectations. It also serves as the gateway to deeper statistical reasoning and analysis in future chapters.

6.2 Types of Probability

Probability can be approached from different philosophical and practical perspectives. While the core idea remains the same—quantifying uncertainty—the methodology by which we assign probabilities can differ widely depending on context, available data, and the nature of the event in question.

There are three primary types or approaches to probability:

Classical Probability (Theoretical Probability)

Classical probability is based on the assumption that all outcomes in the sample space are equally likely to occur. It is often used when dealing with games of chance or controlled environments where each possible result has the same likelihood.

Formula:

$$P(A) = \frac{\text{Number of favorable outcomes}}{\text{Total number of possible outcomes}}$$

Example: If you roll a fair six-sided die, there are six equally likely outcomes: $\{1, 2, 3, 4, 5, 6\}$. The probability of rolling a 2 is:

$$P(2) = \frac{1}{6}$$

Another Example: The probability of drawing a heart from a standard deck of 52 cards:

$$P(\text{Heart}) = \frac{13}{52} = \frac{1}{4}$$

Key Features:

- Requires a finite sample space.
- Assumes outcomes are equally likely.
- Used in ideal, controlled, or hypothetical scenarios.

Limitations: In the real world, not all outcomes are equally likely. For example, assuming equal likelihood in biased games, real-world experiments, or unpredictable systems may lead to inaccurate conclusions.

Empirical Probability (Relative Frequency)

Empirical probability, also known as **experimental probability**, is based on direct observation or historical records. Instead of relying on theory, we compute probabilities from actual data by looking at how often an event occurs over many trials.

Formula:

$$P(A) = \frac{\text{Number of times event A occurred}}{\text{Total number of trials}}$$

Example: NASA has flown 113 space shuttle missions. If 111 were successful, the empirical probability of a successful mission is:

$$P(\text{Success}) = \frac{111}{113} \approx 0.982$$

Another Example: A quality control engineer inspects 1,000 lightbulbs and finds 28 are defective. The empirical probability that a lightbulb is defective is:

$$P(\text{Defective}) = \frac{28}{1000} = 0.028$$

Key Features:

- Based on data and actual experience.
- Can update over time as more data is collected.

- Useful when theoretical models are not feasible.

Limitations: If the number of trials is too small or not representative of the full population, the estimate may be inaccurate. For rare events, very large datasets are needed to obtain meaningful probabilities.

Law of Large Numbers: Over a large number of trials, empirical probability tends to converge to the theoretical (classical) probability. This principle underlies the reliability of long-term data analysis.

6.2.1 Subjective Probability

Subjective probability reflects personal belief, intuition, or estimation about how likely an event is to occur. It is used when no historical data or equally likely outcomes are available, such as in speculative or unique events.

Example: Estimating the probability that a startup will become a billion-dollar company in the next five years. There may be no concrete data to reference, so the estimate is based on the investor's intuition, experience, or optimism.

Another Example: A meteorologist might say there's a 60% chance of rain tomorrow based not just on models, but also on years of personal judgment interpreting atmospheric patterns.

Key Features:

- Based on human judgment, feelings, or expert opinion.

- Often used in business, politics, law, and other decision-making under uncertainty.
- May incorporate personal knowledge, experience, or even gut feeling.

Limitations: Subjective probability is prone to bias, inconsistency, and emotional influence. Two people can assign very different probabilities to the same event depending on their perspectives or information.

Example in Court: A juror might believe there's a 90% chance the defendant is guilty, even without statistical proof. This reflects a subjective interpretation of the evidence.

Summary Comparison Table

Type	Definition	Example
Classical	Based on equally likely outcomes in a finite sample space	Rolling a die: $P(4) = \frac{1}{6}$
Empirical	Based on observed data or experimentation	111 out of 113 NASA missions succeeded: $P(\text{Success}) = \frac{111}{113}$
Subjective	Based on personal judgment or opinion	"I believe there's an 80% chance that this project will succeed."

Why These Types Matter

Understanding these three approaches helps us choose the right method for different situations. In academic experiments, classical probability may be most appropriate. In business or scientific research, empirical methods rule. In uncertain scenarios, like predicting market crashes or geopolitical outcomes, subjective probability can offer useful—even if imperfect—insight.

Together, these types form the foundation of how probability is used across fields like economics, engineering, computer science, medicine, and psychology.

6.3 Basic Terminology

- **Experiment:** A process that leads to a single outcome.
- **Outcome:** The result of an experiment.
- **Event:** A collection of one or more outcomes.
- **Mutually Exclusive Events:** Events that cannot happen at the same time.
- **Collectively Exhaustive Events:** At least one of the events must occur.
- **Independent Events:** The outcome of one event does not affect the other.

6.4 Rules of Probability

Understanding the rules of probability helps us quantify and analyze the likelihood of various outcomes. Whether we're assessing risk, planning strategies, or interpreting statistical data, these foundational rules allow us to combine, contrast, and reason about the likelihood of complex events.

6.4.1 Addition Rule

The addition rule helps calculate the probability that *at least one* of two events will occur. However, how we apply the rule depends on whether the events are mutually exclusive.

Special Rule (Mutually Exclusive Events)

Two events are **mutually exclusive** if they cannot occur at the same time. That is, the occurrence of one event precludes the occurrence of the other. If A and B are mutually exclusive, then:

$$P(A \cup B) = P(A) + P(B)$$

Example: When flipping a fair coin:

- Let A be the event of getting heads: $P(A) = 0.5$
- Let B be the event of getting tails: $P(B) = 0.5$
- Since a coin cannot land on both heads and tails

simultaneously, A and B are mutually exclusive.

$$P(\text{Heads or Tails}) = 0.5 + 0.5 = 1$$

General Rule (Not Mutually Exclusive)

If two events **can** occur at the same time, they are not mutually exclusive. In this case, adding their probabilities would double-count the outcomes they share. To correct for this, we subtract the probability of their intersection:

$$P(A \cup B) = P(A) + P(B) - P(A \cap B)$$

Example: Suppose in a class of 30 students:

- 18 study math ($P(M) = \frac{18}{30}$),
- 12 study science ($P(S) = \frac{12}{30}$),
- 6 study both ($P(M \cap S) = \frac{6}{30}$).

Then the probability a randomly chosen student studies math or science is:

$$P(M \cup S) = \frac{18}{30} + \frac{12}{30} - \frac{6}{30} = \frac{24}{30} = 0.8$$

Interpretation: There's an 80% chance a randomly chosen student studies math or science.

6.4.2 Complement Rule

The **complement** of an event A, denoted $\bar{A}$ or "not A," represents all outcomes where A does not occur. Since either A occurs or it doesn't, their probabilities must add up to 1:

$$P(A) = 1 - P(\text{not } A)$$

Example: If there's a 70% chance of rain tomorrow:

$$P(\text{No Rain}) = 1 - 0.70 = 0.30$$

Why These Rules Matter

- The **addition rule** helps assess the probability of multiple opportunities or risks occurring.
- The **complement rule** simplifies problems—often it's easier to calculate what you don't want and subtract from 1.
- These rules underpin probability trees, Venn diagrams, and decision-making tools in business, science, and everyday life.

Real-World Application: In cybersecurity, the addition rule helps estimate risk when multiple vulnerabilities exist. In insurance, the complement rule helps assess "survival" probabilities in actuarial tables. In marketing, it's used to measure campaign overlaps and responses.

Always ask:

1. Are these events able to happen at the same time? (If yes, they are not mutually exclusive.)
2. Would it be easier to find the complement? (E.g., instead of calculating "at least one," try subtracting the probability of "none.")

6.5 Multiplication Rule

The multiplication rule allows us to determine the probability that two events happen together (in sequence or at the same time), denoted as $P(A \cap B)$. Understanding this rule helps us handle compound probabilities and joint likelihoods in both simple and complex scenarios.

There are two cases to consider: when events are **independent** and when they are **dependent**. The distinction is crucial in deciding which version of the multiplication rule to apply.

6.5.1 Independent Events

Two events are said to be **independent** if the outcome of one does not affect the probability of the other. In such a case, the probability that both events occur is simply the product of their individual probabilities:

$$P(A \cap B) = P(A) \cdot P(B)$$

Example: Suppose we roll a fair six-sided die and flip a fair coin.

- Let A be the event of rolling a 3: $P(A) = \frac{1}{6}$
- Let B be the event of flipping a heads: $P(B) = \frac{1}{2}$

Since rolling a die and flipping a coin don't affect each other, they are independent:

$$P(A \cap B) = \frac{1}{6} \cdot \frac{1}{2} = \frac{1}{12}$$

Interpretation: There is an 8.33% chance of rolling a 3 and flipping a heads.

Real-world application: Independent events are common in simulations and probabilistic modeling—for example, determining the likelihood of independent equipment failures or parallel tasks succeeding in project management.

6.5.2 Dependent Events

Events are **dependent** if the outcome or occurrence of the first event affects the probability of the second. In these cases, we must adjust our calculation using conditional probability. The multiplication rule becomes:

$$P(A \cap B) = P(A) \cdot P(B|A)$$

Where $P(B|A)$ is the probability of event B occurring

given that A has already occurred.

Example: Suppose we draw two cards from a standard deck of 52 cards *without replacement.*

- Let A be the event that the first card drawn is an Ace.
- Let B be the event that the second card drawn is also an Ace.

$$P(A) = \frac{4}{52}, \quad P(B|A) = \frac{3}{51}$$

Since the deck changes after the first draw, the events are dependent:

$$P(A \cap B) = \frac{4}{52} \cdot \frac{3}{51} = \frac{12}{2652} \approx 0.0045$$

Interpretation: There is a 0.45% chance of drawing two Aces in a row without replacement.

Why This Matters:

- **Independent Events** reflect scenarios where each trial resets conditions—ideal for modeling games of chance, physical systems, and simulations.
- **Dependent Events** occur frequently in the real world, especially where limited resources, memory effects, or sequences exist (e.g., medical trials, quality control, or sequential sampling).

Common Mistake to Avoid: Do not assume independence when the outcome of one event affects the other. For instance, in surveys where participants are not replaced or in events involving timing or causality, dependence must be assumed.

Quick Summary Table

Scenario	Use This Formula
Independent Events	$P(A \cap B) = P(A) \cdot P(B)$
Dependent Events	$P(A \cap B) = P(A) \cdot P(B \mid A)$

Mnemonic Tip: *"If A affects B, adjust accordingly. If not, just multiply directly."*

6.6 Conditional Probability

Conditional probability refers to the likelihood of an event occurring given that another event has already taken place. It allows us to update our expectations based on new information.

$$P(A|B) = \frac{P(A \cap B)}{P(B)} \quad \text{provided } P(B) > 0$$

Explanation: This formula expresses the probability of event A happening under the condition that B has already occurred. It is read as "the probability of A given B." This concept is essential in everyday reasoning, data science, medical diagnostics, legal inference, and more.

Intuitive Interpretation

Imagine you're in a town where 20% of the population are children, and 40% of the children wear glasses. The probability that a person is a child *and* wears glasses is:

$$P(\text{Child} \cap \text{Glasses}) = 0.20 \cdot 0.40 = 0.08$$

But if you already know the person is a child, the probability they wear glasses becomes:

$$P(\text{Glasses}|\text{Child}) = \frac{P(\text{Child} \cap \text{Glasses})}{P(\text{Child})} = \frac{0.08}{0.20} = 0.40$$

This shows that conditional probability helps you refine your expectations based on context.

Real-World Example: Medical Testing

Suppose 1% of a population has a certain disease. A test correctly detects the disease 99% of the time (true positive rate) and has a 5% false positive rate. If a person tests positive, what is the probability they actually have the disease?

Let:

- D = Has disease
- T = Tests positive

We are looking for:

$$P(D|T) = \frac{P(D \cap T)}{P(T)}$$

Assuming we test 10,000 people:

- People with disease: $0.01 \cdot 10,000 = 100$
- True positives: $0.99 \cdot 100 = 99$
- False positives: $0.05 \cdot 9,900 = 495$
- Total testing positive: $99 + 495 = 594$

So:

$$P(D|T) = \frac{99}{594} \approx 0.167$$

Surprise: Even though the test is 99% accurate, a positive result only gives a 16.7% chance of actually having the disease. This is why context and base rates are critical—one of the most important applications of conditional probability.

Dependent vs. Independent Context

If A and B are independent:

$$P(A|B) = P(A)$$

In other words, knowing that B occurred does not change the probability of A occurring. This serves as a test for independence.

If A and B are dependent:

$$P(A|B) \neq P(A)$$

Meaning the occurrence of B provides new information that shifts how we evaluate A.

Tree Diagram Visualization

A probability tree can help visualize conditional events:

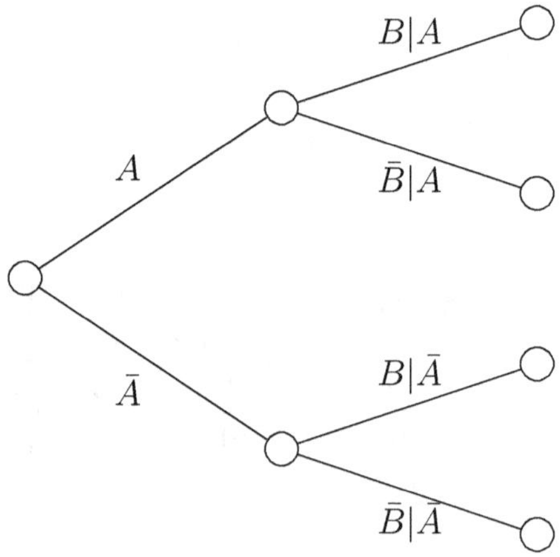

Each branch represents a sequential path and its associated conditional probability. Multiplying along a branch gives the joint probability of the full event sequence.

Common Pitfalls

- **Assuming Symmetry:** $P(A|B)$ is not equal to $P(B|A)$. Confusing these leads to serious reasoning errors, especially in medicine, law, and policy.

- **Ignoring Base Rates:** When focusing only on conditional accuracy (e.g., 99% test reliability), ignoring the base rate (e.g., only 1% prevalence) leads to misleading conclusions.
- **Forgetting the Denominator:** Conditional probability always divides by $P(B)$, not by the total sample space.

Applications of Conditional Probability

- **Bayesian Statistics:** Updates beliefs or estimates as more data is observed.
- **Machine Learning:** Algorithms like Naive Bayes rely on conditional independence assumptions.
- **Healthcare:** Risk assessments and decision trees use conditional logic.
- **Finance:** Conditional Value at Risk (CVaR) estimates potential losses.
- **Criminology and Law:** Legal reasoning and evidence evaluation often involve conditional logic.

Final Thought

Conditional probability allows us to make informed predictions and decisions based on changing circumstances. In an uncertain world, updating our expectations with context is not just useful—it's necessary.

6.7 Bayes' Theorem

Bayes' Theorem is a powerful and foundational principle in probability theory that allows us to update the probability of an event based on new information or evidence. It bridges the gap between theoretical probability and observed data, offering a structured way to reason under uncertainty.

$$P(A|B) = \frac{P(B|A) \cdot P(A)}{P(B)} \quad \text{provided } P(B) > 0$$

This formula gives us the probability of event A occurring given that event B has occurred. It reverses the conditionality—starting with what we know about B given A and transforming it into a statement about A given B.

Terminology Breakdown

- $P(A)$: Prior probability – our initial belief about A before seeing new evidence.
- $P(B|A)$: Likelihood – the probability of observing B assuming A is true.
- $P(B)$: Marginal probability – the total probability

of observing B from all sources.

- $P(A|B)$: Posterior probability – the revised belief about A after considering B.

Why Bayes' Theorem Matters

Bayes' Theorem is not just a mathematical curiosity; it is the foundation of *Bayesian thinking*, which contrasts with classical (frequentist) probability. Instead of just recording outcomes and frequencies, Bayesian probability is about belief revision — how we update our beliefs as we gain new evidence.

Real-Life Example: Medical Diagnosis

Suppose:

- 1% of people in a city have a rare disease ($P(D) = 0.01$).
- A diagnostic test correctly identifies the disease 95% of the time ($P(\text{Positive}|D) = 0.95$).
- The test gives false positives 2% of the time ($P(\text{Positive}|\neg D) = 0.02$).

What is the probability that a person who tested positive actually has the disease? That is, what is $P(D|\text{Positive})$?

Step 1: Use Bayes' Theorem

$$P(D|\text{Positive}) = \frac{P(\text{Positive}|D) \cdot P(D)}{P(\text{Positive})}$$

We need to compute $P(\text{Positive})$ using the law of total probability:

$$P(\text{Positive}) = P(\text{Positive}|D) \cdot P(D) + P(\text{Positive}|\neg D) \cdot P(\neg D)$$

$$= 0.95 \cdot 0.01 + 0.02 \cdot 0.99 = 0.0095 + 0.0198 = 0.0293$$

Step 2: Plug into Bayes' Theorem

$$P(D|\text{Positive}) = \frac{0.95 \cdot 0.01}{0.0293} = \frac{0.0095}{0.0293} \approx 0.324$$

Interpretation: Even with a 95% accurate test, a person who tests positive only has about a 32.4% chance of actually having the disease! This reveals the importance of base rates and context.

Bayesian Thinking in Everyday Life

Bayes' Theorem is essential in:

- **Spam Filtering:** Email systems use Bayes' Theorem to decide whether a message is spam based on word patterns.
- **Medical Testing:** Doctors use Bayesian reasoning to update a diagnosis after each test.

- **Legal Reasoning:** Juries might (implicitly) apply Bayesian updates when new evidence is presented during a trial.
- **Machine Learning:** Algorithms like Naive Bayes Classifier rely directly on this theorem.
- **Weather Forecasting:** Meteorologists refine predictions as new atmospheric data is observed.

Bayes' Theorem Table Format (Optional)

	Has Disease (D)	No Disease ($\neg D$)	Total
Test Positive	95	198	293
Test Negative	5	9,702	9,707
Total	100	9,900	10,000

Table 6.1: Example: Diagnostic Test Results in a Population of 10,000

$$P(D|\text{Positive}) = \frac{95}{293} \approx 0.324$$

This format helps visualize the real-world meaning of Bayesian analysis.

The Logic of Reversal

Many people intuitively confuse $P(A|B)$ with $P(B|A)$. Bayes' Theorem provides a way to reverse conditional probabilities when the direct version is difficult to observe.

Example: If you know how likely you are to hear thunder when there's lightning, Bayes lets you infer how likely it is there was lightning when you heard thunder.

Generalized Bayes' Theorem (Multiple Hypotheses)

$$P(H_i|E) = \frac{P(E|H_i) \cdot P(H_i)}{\sum_{j=1}^{n} P(E|H_j) \cdot P(H_j)}$$

Where:

- H_i = one of several competing hypotheses.
- E = observed evidence.
- The denominator is the total probability of the evidence across all hypotheses.

Conclusion

Bayes' Theorem is more than a formula—it's a mindset. It teaches us to:

- Begin with what we believe (prior),
- Integrate new evidence (likelihood),
- Reevaluate our position (posterior).

In a world where information is constantly updated, Bayesian thinking empowers better decision-making and more honest self-correction.

6.8 Joint Probability

Joint probability refers to the likelihood of two events occurring simultaneously. It is denoted as $P(A \cap B)$, representing the probability that both event A and event B occur. This concept is especially useful when analyzing relationships between categorical variables, and it lays the foundation for understanding more advanced statistical tools like correlation, regression, and Bayesian inference.

Understanding Joint Probability Through Contingency Tables

Contingency tables (also called cross-tabulations or two-way tables) are valuable tools for organizing and displaying the frequency of combinations of categories. Each cell in the table represents the joint frequency of two categories occurring together, and by dividing each frequency by the grand total, we get the corresponding joint probabilities.

Example: Employee Loyalty and Years of Service

Consider a company that surveys its employees about their loyalty (Loyal vs. Not Loyal) and their years of service (Less than 10 years vs. 10+ years). The data is presented in a contingency table below:

	< 10 Years	**10+ Years**	**Total**
Loyal	120	180	300
Not Loyal	200	100	300
Total	320	280	600

Table 6.2: Employee Loyalty by Years of Service

Calculating Joint Probabilities

To compute the joint probability of two categories, divide the count in the intersecting cell by the grand total. For instance:

- $P(\text{Loyal} \cap 10+ \text{ Years}) = \frac{180}{600} = 0.30$
- $P(\text{Not Loyal} \cap < 10 \text{ Years}) = \frac{200}{600} \approx 0.333$

These values indicate that 30

Why Joint Probability Matters

Understanding joint probability helps businesses and analysts:

- Discover patterns and associations between different categorical variables.
- Evaluate the likelihood of intersecting traits (e.g., loyalty and experience).
- Assess risk by analyzing the simultaneous occurrence of multiple factors (e.g., product failure and high cost).

Joint probabilities are particularly useful in market research, healthcare analytics, employee retention strategies, quality control, and risk management. For example, in epidemiology, researchers might calculate the joint probability of a patient being both a smoker and having lung disease. In finance, an analyst may look at the joint likelihood of a stock increasing in price while interest rates decrease.

Connection to Other Probability Concepts

Joint probability also serves as the foundation for:

- **Conditional Probability:** How one event affects the likelihood of another.
- **Bayes' Theorem:** Updating beliefs based on new data.
- **Multiplication Rule:** Computing joint probabilities of dependent and independent events.

Visualizing Joint Probability

Joint probabilities are often visualized using:

- **Venn Diagrams** for overlapping categories.
- **Probability Trees** for sequential events.
- **Heat Maps** and **Mosaic Plots** in data visualization software to depict densities and frequencies.

Summary

Joint probability expands our understanding of how variables interact and co-occur. By leveraging contingency tables and basic formulas, analysts can quickly draw insights about relationships and dependencies that may otherwise go unnoticed. Whether in academia, business, or everyday decision-making, joint probability provides a crucial lens for understanding multi-variable situations.

6.9 Tree Diagrams

Tree diagrams are powerful visual tools used to map out and calculate probabilities for multi-stage experiments or sequential events. Each branch in the tree represents a possible outcome and is labeled with the probability of that outcome. As you move through stages of the tree from left to right, you multiply the probabilities along each path to determine the joint probability of a specific outcome sequence.

Why Use Tree Diagrams?

Tree diagrams help organize complex probability problems by:

- Displaying all possible outcomes clearly and logically.
- Breaking down compound events into smaller, manageable parts.

- Making it easy to calculate joint, marginal, and conditional probabilities.
- Helping to visualize dependent and independent events across stages.

Basic Structure of a Tree Diagram

A tree diagram starts with a root node (representing the starting point of an experiment). Branches extend out for each possible outcome of the first event. From each of those branches, more branches extend to represent the outcomes of the second event, and so on.

Key Features:

- The sum of probabilities for branches coming out of a single point is always 1.
- Multiply the probabilities along a path to compute the joint probability.
- Add up the probabilities of paths leading to the same final outcome to get marginal probabilities.

Example: Drawing Colored Balls from a Bag (Without Replacement)

Imagine a bag containing 2 red balls (R) and 1 blue ball (B). You draw two balls one after the other without replacement.

Stage 1 (First Draw):

- Probability of R $= \frac{2}{3}$
- Probability of B $= \frac{1}{3}$

Stage 2 (Second Draw, Conditional):

- If R was drawn first: 1 red, 1 blue remain $\rightarrow$ $P(R|R) = \frac{1}{2},\ P(B|R) = \frac{1}{2}$
- If B was drawn first: 2 red remain $\rightarrow$ $P(R|B) = 1,\ P(B|B) = 0$

Tree Diagram Illustration

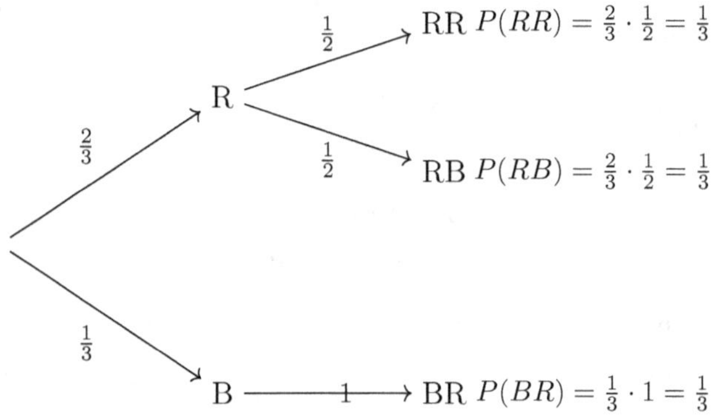

Figure 6.1: Tree Diagram: Drawing Two Balls Without Replacement

Interpreting the Diagram

Each complete path from left to right represents a possible outcome of the two draws:

- RR: Red then Red

- RB: Red then Blue
- BR: Blue then Red

Notice there is no BB outcome, since after removing the only blue ball on the first draw, no blue ball remains for the second.

When to Use Tree Diagrams

Use tree diagrams when:

- You have multiple steps or stages in an experiment.
- You need to differentiate between dependent and independent outcomes.
- You're calculating conditional probabilities.
- Visual clarity is needed to understand all possible outcomes.

Summary

Tree diagrams provide an intuitive and structured way to handle multi-stage probability problems. By breaking down each event and clearly identifying all possible paths, they reduce errors in probability calculations. Whether you're working with replacement, without replacement, or analyzing dependent and independent events, tree diagrams can help you visualize, calculate, and interpret probabilities with clarity and precision.

6.10 Counting Rules

Counting rules form the mathematical foundation for determining how many ways events can occur. These techniques are crucial for calculating probabilities, particularly when the total number of outcomes is large or complex. This section will explore key counting principles, including the multiplication rule, permutations, and combinations.

6.10.1 Multiplication Rule

The multiplication rule states that if a task can be performed in m ways and another independent task can be performed in n ways, then both tasks together can be performed in $m \cdot n$ ways.

Example: Suppose you have 3 different shirts and 2 different pairs of pants. How many outfits can you create?

$$\text{Total outfits} = 3 \times 2 = 6$$

Each shirt can be paired with each of the 2 pants. This rule can be extended to any number of stages.

Generalization: If there are k tasks with $n_1, n_2, ..., n_k$ possible outcomes, then the total number of combinations is:

$$n_1 \times n_2 \times \cdots \times n_k$$

Real-Life Use: This principle is used in creating passwords. If a password consists of 4 digits, and each digit can be from 0 to 9, the number of possible 4-digit combinations is:

$$10 \times 10 \times 10 \times 10 = 10^4 = 10,000$$

6.10.2 Permutations

Permutations refer to the number of ways to arrange r objects out of n total objects, where the order of selection is important.

$$P(n, r) = \frac{n!}{(n-r)!}$$

Example: Suppose there are 5 contestants in a race, and we want to know how many ways we can award gold, silver, and bronze medals (1st, 2nd, and 3rd place).

$$P(5, 3) = \frac{5!}{(5-3)!} = \frac{5 \cdot 4 \cdot 3 \cdot 2 \cdot 1}{2 \cdot 1} = 60$$

Note: Every different arrangement of the same elements counts as a distinct permutation.

When to Use Permutations

Use permutations when:

- You're assigning ranks or positions.
- You're arranging items in a specific sequence.
- The sequence of selection matters.

6.10.3 Combinations

Combinations count the number of ways to choose r objects from a set of n objects when the order of selection does not matter.

$$C(n, r) = \frac{n!}{r!(n-r)!}$$

Example: A committee of 3 members is to be selected from a group of 10 people. How many possible committees can be formed?

$$C(10, 3) = \frac{10!}{3!(10-3)!} = \frac{10 \cdot 9 \cdot 8}{3 \cdot 2 \cdot 1} = 120$$

Unlike permutations, combinations treat each selection as identical regardless of order.

When to Use Combinations

Use combinations when:

- You're selecting groups without regard to order.
- You're forming teams or committees.

- The sequence in which items are selected is irrelevant.

Permutations vs. Combinations

Feature	Permutations	Combinations
Order Matters?	Yes	No
Formula	$\frac{n!}{(n-r)!}$	$\frac{n!}{r!(n-r)!}$
Use Case	Rankings, Codes	Teams, Groups
Example	Positions in a race	Lottery ticket numbers

Table 6.3: Key Differences Between Permutations and Combinations

Additional Example: Counting Ice Cream Sundaes

Suppose you want to build a 3-scoop ice cream sundae using 5 different flavors. If:

- The order *does* matter (e.g., you care which scoop is on top), then:

$$P(5,3) = \frac{5!}{(5-3)!} = 60 \text{ sundaes}$$

- The order *does not* matter, then:

$$C(5,3) = \frac{5!}{3!(5-3)!} = 10 \text{ sundaes}$$

6.10.4 Summary

Counting rules provide essential tools for computing probabilities in both simple and complex scenarios. Whether we are determining how many ways a password can be formed, how teams are chosen, or how rankings are assigned, understanding the distinction between permutations and combinations is crucial. The multiplication rule offers the foundational framework, and the permutation-combination distinction guides us through real-world decision making with precision.

6.11 Conclusion

Probability is a foundational concept in statistics. Understanding how to compute probabilities using classical, empirical, and subjective approaches—as well as mastering the rules of addition, multiplication, and Bayes' Theorem—enables us to model uncertainty and make informed decisions.

Exercises

1. **Classical Probability:** A six-sided die is rolled once. What is the probability of rolling an even number?

2. **Empirical Probability:** A factory records that 40 out of 200 light bulbs produced in a day are defective. Based on this data, what is the probability that a randomly selected bulb is not defective?

3. **Subjective Probability:** An investor believes there's a 70% chance the stock market will rise this quarter based on expert opinions. Classify this type of probability and explain why it's used.

4. **Complement Rule:** If the probability that it rains tomorrow is 0.35, what is the probability that it does not rain?

5. **Addition Rule (Mutually Exclusive):** A card is drawn from a standard deck. What is the probability that it is either a King or a Queen?

6. **Addition Rule (Not Mutually Exclusive):** A student is chosen at random from a class. The probability that the student is left-handed is 0.1, and the probability the student wears glasses is 0.25. The probability that a student is both left-handed and wears glasses is 0.05. What is the probability the student is either left-handed or wears glasses?

7. **Multiplication Rule (Independent Events):** A coin is flipped and a die is rolled. What is the probability of getting heads and a 3?

8. **Conditional Probability:** In a survey, 60% of respondents like chocolate. Among those, 40% also like vanilla. What is the probability that a person likes vanilla given that they like chocolate?

9. **Tree Diagram & Conditional Probability:** A box contains 3 red balls and 2 blue balls. One ball is drawn at random and not replaced. Then a second ball is drawn. What is the probability that

both balls are red?

10. **Bayes' Theorem:** A medical test is 95% accurate for detecting a disease. The disease occurs in 2% of the population. If a person tests positive, what is the probability that they actually have the disease? (Hint: Use Bayes' Theorem)

Chapter 7

Discrete Distributions

Introduction to Discrete Probability Distributions

A discrete probability distribution lists each possible value a discrete random variable can take, along with its probability. These distributions are foundational in modeling count-based outcomes in engineering, science, economics, and social studies. Discrete variables are those that take on a countable number of distinct values, such as the number of heads when flipping coins, the number of defective items in a batch, or the number of emails received per hour.

Let X be a discrete random variable. Then, the probability distribution of X is a function $P(X = x)$ which assigns a probability to each possible outcome x.

Properties

A discrete probability distribution describes the probability of each possible value a discrete random variable can take. It must satisfy two fundamental mathematical properties, which ensure logical consistency and totality of outcomes:

1. **Bounded Probability:** $0 \leq P(X = x_i) \leq 1$ for every value x_i in the sample space. This means the probability of any single outcome must be at least 0 (impossible) and at most 1 (certain). No probability can be negative or exceed 1. This rule ensures realism in our models—no outcome can be "more than guaranteed" or "less than impossible."
2. **Total Probability Must Equal 1:** $\sum_i P(X = x_i) = 1$. The sum of the probabilities of all possible outcomes must equal 1. This ensures that when considering the entire sample space, we have accounted for all possibilities—something must happen. Any missing probability would suggest that some outcome was left out of consideration, which violates the completeness of the distribution.

Example 1: Rolling a Fair Die

Let X represent the result of rolling a fair six-sided die. The possible values are $1, 2, 3, 4, 5, 6$, and each has equal probability:

$$P(X = x) = \frac{1}{6}, \quad \text{for } x = 1, 2, 3, 4, 5, 6$$

- Each probability is between 0 and 1.
- The total probability is:

$$\sum_{x=1}^{6} \frac{1}{6} = \frac{6}{6} = 1$$

This satisfies both properties and is therefore a valid discrete probability distribution.

Example 2: Biased Coin

Let X represent the result of flipping a biased coin where the probability of heads is 0.7 and tails is 0.3:

$$P(X = \text{Heads}) = 0.7, \quad P(X = \text{Tails}) = 0.3$$

- Each value is between 0 and 1.
- The total probability is:

$$0.7 + 0.3 = 1$$

This distribution also meets the required criteria.

Why It Matters

These properties ensure mathematical validity and statistical soundness. If either rule is violated, we cannot call the function a legitimate probability distribution. For instance, suppose a function gave

$P(X = 2) = 1.1$—this would be invalid, as it violates the basic rule that probabilities cannot exceed 1.

Likewise, if the sum of all probabilities was 0.97, it would suggest that we forgot to account for a possible outcome—or made a miscalculation. This could lead to faulty models, wrong predictions, and potentially costly real-world decisions.

Understanding and checking these properties is foundational to working confidently with any discrete random variable.

7.1 PMF

A **Probability Mass Function (PMF)** provides the complete specification of a discrete probability distribution. It connects outcomes to their probabilities in a precise and measurable way.

Definition

Let X be a discrete random variable. The PMF is a function $p(x)$ such that:

- $p(x) = P(X = x)$, which is the probability that the random variable X equals a specific value x.
- The function must satisfy:

$$\sum_x p(x) = 1$$

This means that the total probability across all possible values of X must sum to 1.

Important Notes

- A PMF is only defined for discrete random variables—those which can take on a countable number of distinct values (e.g., 0, 1, 2, 3).
- $p(x) \geq 0$ for every x, and the PMF is 0 for any x not in the sample space.
- The PMF provides a map from each outcome to its likelihood.

Motivation and Usefulness

The PMF gives us a compact, mathematical way to represent uncertainty. With it, we can:

- Calculate the probability of any particular outcome.
- Compute expected values, variances, and other statistics.
- Compare different distributions or scenarios.
- Build models for games of chance, machine reliability, queueing systems, and more.

Example: Coin Tosses

Let's say we flip a fair coin 3 times. Let the random variable X represent the number of heads observed. The possible outcomes in the sample space are:

$$\{TTT, TTH, THT, HTT, HHT, HTH, THH, HHH\}$$

Each sequence of tosses has an equal probability of $\frac{1}{8}$ because there are 8 total equally likely outcomes (2 outcomes per toss, $2^3 = 8$).

Let's find the possible values of X:

$$X \in \{0, 1, 2, 3\}$$

And now the PMF:

$$\begin{aligned}
P(X = 0) &= \frac{1}{8} \quad \text{(Only TTT)} \\
P(X = 1) &= \frac{3}{8} \quad \text{(TTH, THT, HTT)} \\
P(X = 2) &= \frac{3}{8} \quad \text{(HHT, HTH, THH)} \\
P(X = 3) &= \frac{1}{8} \quad \text{(Only HHH)}
\end{aligned}$$

Number of Heads, X	Probability, $P(X)$
0	$\frac{1}{8}$
1	$\frac{3}{8}$
2	$\frac{3}{8}$
3	$\frac{1}{8}$

Graphical Representation

It is common to plot the PMF using a bar graph, where the x-axis shows the values of the random variable and the y-axis shows the probability of each value. The heights of the bars correspond to the values of the PMF.

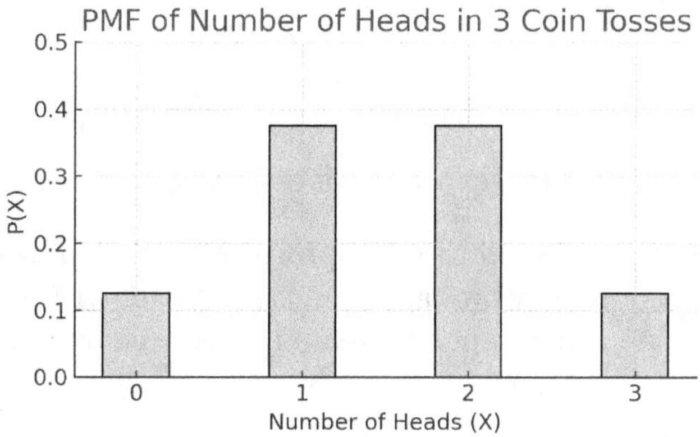

Figure 7.1: Probability Mass Function for Number of Heads in 3 Coin Tosses

As shown in Figure 7.1, this bar graph clearly illustrates the symmetric distribution of outcomes for 3 coin tosses. The most likely outcomes are 1 or 2 heads, with probabilities peaking at those values. This type of visualization makes it easy to compare relative likelihoods across discrete outcomes.

Interpretation

This PMF tells us, for example:

- The most likely outcomes are 1 or 2 heads, each with probability $\frac{3}{8}$.
- The extremes (0 or 3 heads) are less likely, each with probability $\frac{1}{8}$.
- This distribution is symmetric because the coin is fair.

Computing Mean, Variance, and Standard Deviation of a PMF

When working with discrete probability distributions, especially when a probability mass function (PMF) is known, we can compute three important characteristics of the distribution:

- **Mean (Expected Value)** — the long-run average of outcomes
- **Variance** — the average squared deviation from the mean
- **Standard Deviation** — the square root of the variance

Step-by-Step Method

Given a discrete random variable X with PMF $P(X = x_i)$:

1. Compute the **mean**:

$$\mu = E(X) = \sum x_i \cdot P(x_i)$$

2. Compute the **variance**:

$$\sigma^2 = \sum x_i^2 \cdot P(x_i) - \mu^2$$

3. Compute the **standard deviation**:

$$\sigma = \sqrt{\sigma^2}$$

Example: Tossing a Coin 3 Times

Let X be the number of heads in 3 tosses of a fair coin. The sample space is:

$$\{TTT, TTH, THT, HTT, HHT, HTH, THH, HHH\}$$

So the values of X and their probabilities are:

X	$P(X)$
0	$\frac{1}{8}$
1	$\frac{3}{8}$
2	$\frac{3}{8}$
3	$\frac{1}{8}$

Step 1: Compute the Expected Value or Mean

$$\mu = 0\cdot\frac{1}{8}+1\cdot\frac{3}{8}+2\cdot\frac{3}{8}+3\cdot\frac{1}{8} = 0+\frac{3}{8}+\frac{6}{8}+\frac{3}{8} = \frac{12}{8} = 1.5$$

Step 2: Compute the Variance

$$\sum x_i^2 \cdot P(x_i) = 0^2\cdot\frac{1}{8}+1^2\cdot\frac{3}{8}+2^2\cdot\frac{3}{8}+3^2\cdot\frac{1}{8} = 0+\frac{3}{8}+\frac{12}{8}+\frac{9}{8} = \frac{24}{8} = 3$$

$$\sigma^2 = 3 - (1.5)^2 = 3 - 2.25 = 0.75$$

Step 3: Compute the Standard Deviation

$$\sigma = \sqrt{0.75} \approx 0.866$$

Interpretation

The expected number of heads is 1.5. The standard deviation of approximately 0.866 tells us how much actual outcomes are likely to vary from that average. Since this is a symmetric and fairly narrow distribution, the values stay close to the center.

Common Pitfall

Students often confuse a PMF with a cumulative distribution function (CDF). Remember:

- The PMF tells you the probability of a single value.
- The CDF tells you the probability of the variable being less than or equal to a value.

Why PMFs Matter

Understanding the PMF helps you:

- Quantify and communicate risk.
- Understand games, simulations, and probabilistic models.
- Develop foundational knowledge for studying expected value, variance, and inferential statistics.

A well-defined PMF is your first step toward building and analyzing discrete probability models with mathematical precision.

7.2 Graphical Representation

A *graphical representation* is an essential way to visualize a discrete probability distribution. It allows us to intuitively interpret and compare the probabilities associated with different outcomes of a discrete random variable. The most commonly used type of plot for discrete distributions is the bar graph.

Bar Graph Description

In a bar graph representing a **Probability Mass Function (PMF)**:

- The **x-axis** represents the possible values that the discrete random variable X can take.
- The **y-axis** represents the corresponding

probabilities $P(X = x)$.

- Each bar is centered over a value of X and its height reflects the probability associated with that value.
- The bars are not connected, emphasizing that the variable is discrete (i.e., it jumps from one value to another with no values in between).

Visual Example

Let's revisit the coin-toss experiment where X is the number of heads observed in 3 tosses of a fair coin. We previously determined:

X	$P(X)$
0	$\frac{1}{8}$
1	$\frac{3}{8}$
2	$\frac{3}{8}$
3	$\frac{1}{8}$

The graphical representation of this PMF would look like a bar graph with 4 bars, each centered at $X = 0, 1, 2, 3$, and heights corresponding to $\frac{1}{8}, \frac{3}{8}, \frac{3}{8}, \frac{1}{8}$ respectively.

Why It Matters

- Graphs allow for **quick comparisons**—you can instantly see which outcomes are more likely.

- They help identify **symmetry or skewness** in the distribution.
- They provide visual intuition about how probabilities are distributed, which complements the numerical understanding.
- In teaching and communication, graphs help convey abstract probability ideas more clearly.

Real-Life Example

Suppose a quality control inspector records the number of defective items in randomly selected boxes of 10 products, with probabilities calculated from historical data. A bar graph of these probabilities would quickly show whether most boxes have 0–1 defects (indicating good quality) or higher numbers (suggesting a problem in the process).

Key Takeaway

While tables and formulas are mathematically precise, visualization makes probability distributions accessible. Being able to interpret a bar graph not only aids understanding, but also helps spot inconsistencies, anomalies, and trends in the data.

Tip: If probabilities do not sum to 1, the graph will look "off." Always verify the PMF before graphing!

7.3 CDF

The **Cumulative Distribution Function (CDF)** of a discrete random variable X gives the probability that X will take a value less than or equal to x. That is:

$$F(x) = P(X \leq x) = \sum_{t \leq x} P(X = t)$$

The CDF adds up the probabilities from the probability mass function (PMF) up to a certain value.

Properties of the CDF

- $0 \leq F(x) \leq 1$ for all x
- $F(x)$ is a non-decreasing function
- $\lim_{x \to -\infty} F(x) = 0$ and $\lim_{x \to \infty} F(x) = 1$

Example: CDF from PMF of Coin Tosses

Let X be the number of heads when tossing a fair coin three times. The PMF of X is:

X	$P(X)$
0	$\frac{1}{8}$
1	$\frac{3}{8}$
2	$\frac{3}{8}$
3	$\frac{1}{8}$

Now we compute the cumulative probabilities:

x	$F(x) = P(X \leq x)$
0	$\frac{1}{8}$
1	$\frac{1}{8} + \frac{3}{8} = \frac{4}{8}$
2	$\frac{4}{8} + \frac{3}{8} = \frac{7}{8}$
3	$\frac{7}{8} + \frac{1}{8} = 1$

So the CDF values are:

$$F(0) = \frac{1}{8}, \quad F(1) = \frac{4}{8}, \quad F(2) = \frac{7}{8}, \quad F(3) = 1$$

Graphical Representation

The CDF can be plotted as a step function. Each "step" increases by the value of the PMF at each x.

- At $x = 0$, $F(0) = 0.125$
- At $x = 1$, $F(1) = 0.5$
- At $x = 2$, $F(2) = 0.875$
- At $x = 3$, $F(3) = 1$

Why the CDF Matters

The CDF allows us to compute probabilities over intervals:

$$P(a \leq X \leq b) = F(b) - F(a - 1)$$

For example, the probability of getting between 1 and 2

heads:

$$P(1 \leq X \leq 2) = F(2) - F(0) = \frac{7}{8} - \frac{1}{8} = \frac{6}{8} = 0.75$$

This makes the CDF a valuable tool in summarizing and computing probabilities over a range of values.

7.4 Binomial Distribution

7.4.1 Definition

The binomial distribution is one of the most important and widely used discrete probability distributions in statistics. It describes the number of successes in a fixed number of independent trials, where each trial results in one of only two possible outcomes: success or failure. These outcomes are often labeled as "yes/no," "pass/fail," or "win/lose."

- **Fixed number of trials (n):** The number of times an experiment is repeated is predetermined.
- **Two outcomes:** Each trial results in either success or failure.
- **Constant probability of success (p):** The likelihood of success remains unchanged from trial to trial.
- **Independent trials:** The outcome of one trial does not affect the outcome of another.

This distribution applies in a wide variety of real-life

situations. For example:

- Counting how many lightbulbs out of 20 work.
- Determining how many customers out of 100 buy a product.
- Finding how many voters in a sample of 50 support a specific candidate.

7.4.2 Formula

The probability of obtaining exactly x successes in n trials is given by the binomial probability formula:

$$P(X = x) = \binom{n}{x} p^x (1-p)^{n-x}$$

Where:

- n = number of trials
- x = number of successes
- p = probability of success
- $(1-p)$ = probability of failure
- $\binom{n}{x}$ = number of combinations of n items taken x at a time (also written as "n choose x")

7.4.3 Example 1: Coin Toss

Toss a fair coin 3 times. What is the probability of getting exactly 2 heads?

$$P(X = 2) = \binom{3}{2}\left(\frac{1}{2}\right)^2\left(\frac{1}{2}\right)^1 = 3 \cdot \frac{1}{4} \cdot \frac{1}{2} = \frac{3}{8}$$

7.4.4 Example 2: Lightbulb Defects

Suppose 10 lightbulbs are tested, and the probability of finding a defective bulb is 0.1. What is the probability that exactly 2 bulbs are defective?

$$P(X = 2) = \binom{10}{2}(0.1)^2(0.9)^8 = 45 \cdot 0.01 \cdot 0.43046721 \approx 0.1937$$

7.4.5 Mean, Variance, and Standard Deviation

The mean and variability of a binomial distribution can be calculated using the following formulas:

$$\mu = n \cdot p, \quad \sigma^2 = n \cdot p \cdot (1 - p), \quad \sigma = \sqrt{n \cdot p \cdot (1 - p)}$$

Where:

- μ is the expected value (mean)
- σ^2 is the variance
- σ is the standard deviation

Example: If $n = 10$ and $p = 0.3$:

$$\mu = 10 \cdot 0.3 = 3, \quad \sigma^2 = 10 \cdot 0.3 \cdot 0.7 = 2.1, \quad \sigma \approx 1.449$$

7.4.6 Applications

The binomial distribution is used in quality control, survey analysis, risk assessment, and many areas where counting the number of successes in a sequence of trials is important. For example:

- What is the chance that 7 out of 10 people like your product?
- What's the likelihood that fewer than 2 systems fail out of 20?
- What's the probability of getting 5 or more defective parts in a batch?

Visual Aid: Binomial Distribution Histogram

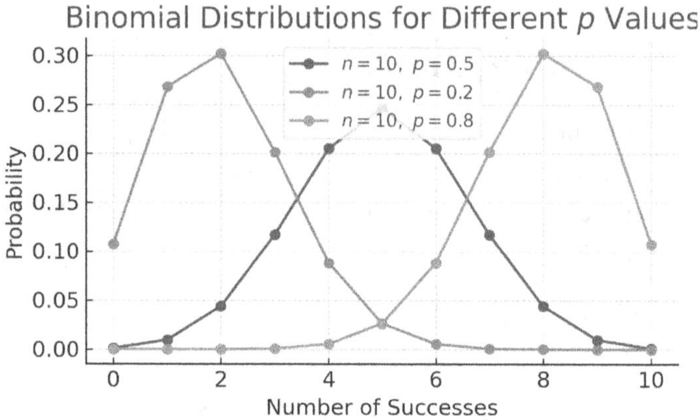

Figure 7.2: Binomial distributions for different values of p and n.

This visual aid shows how the shape of the binomial distribution depends on both n and p:

- When $p = 0.5$, the distribution is symmetric.
- When $p < 0.5$, the distribution is skewed right.
- When $p > 0.5$, the distribution is skewed left.
- As n increases, the distribution becomes more bell-shaped (normal-like).

Tip: When n is large and p is not too close to 0 or 1, the binomial distribution approximates the normal distribution.

7.5 Poisson Distribution

The **Poisson distribution** is a discrete probability distribution used to model the number of times an event occurs in a fixed interval of time, distance, area, or volume when the events happen independently and at a constant average rate. It is particularly useful when modeling rare or infrequent events.

When to Use the Poisson Distribution

The Poisson distribution is appropriate under the following conditions:

- Events occur one at a time.
- Events are independent of each other.
- The average rate (mean number of events per interval), denoted λ, is constant.
- Two events cannot occur at exactly the same instant.

Probability Mass Function (PMF)

The Poisson probability of observing exactly x events in a fixed interval is given by the formula:

$$P(X = x) = \frac{e^{-\lambda}\lambda^x}{x!}$$

where:

- X is the random variable representing the number of occurrences,
- λ is the average number of occurrences in the interval (also the mean and variance),
- e is Euler's number, approximately 2.71828,
- $x!$ is the factorial of x.

Mean and Variance

One of the interesting properties of the Poisson distribution is that its mean and variance are equal:

$$\mu = \lambda \qquad \sigma^2 = \lambda \qquad \sigma = \sqrt{\lambda}$$

Example 1: Website Hits

A website gets 4 hits per minute on average. What is the probability that the site receives exactly 2 hits in a given minute?

$$\begin{aligned} P(X = 2) &= \frac{e^{-4} \cdot 4^2}{2!} \\ &= \frac{e^{-4} \cdot 16}{2} \\ &= 8e^{-4} \\ &\approx 8 \cdot 0.0183 \\ &= 0.1464 \end{aligned}$$

Example 2: Calls at a Call Center

A call center receives an average of 10 calls per hour. What is the probability that exactly 12 calls arrive in an hour?

$$\begin{aligned} P(X = 12) &= \frac{e^{-10} \cdot 10^{12}}{12!} \\ &\approx \frac{0.0000454 \cdot 1{,}000{,}000{,}000{,}000}{479{,}001{,}600} \\ &\approx 0.0948 \end{aligned}$$

Example 3: Misprints in a Book

Suppose on average, there are 2 misprints per 100 pages in a textbook. What is the probability that a randomly selected 100-page section has exactly 0 misprints?

$$\begin{aligned} P(X = 0) &= \frac{e^{-2} \cdot 2^0}{0!} \\ &= e^{-2} \cdot 1 \\ &\approx 0.1353 \end{aligned}$$

Cumulative Poisson Probabilities

To calculate the probability of observing *up to* a certain number of events, sum the individual Poisson probabilities:

$$P(X \leq k) = \sum_{x=0}^{k} \frac{e^{-\lambda}\lambda^x}{x!}$$

This is often done using statistical software or Poisson distribution tables.

Applications of the Poisson Distribution

- Number of phone calls received at a help desk per hour
- Number of accidents at an intersection per month
- Number of decay events from a radioactive source per second
- Number of misprints on a page or in a book
- Number of customer arrivals at a store in a given interval
- Number of spelling errors per chapter in a manuscript

Poisson Approximation to Binomial

When the number of trials n is large and the probability of success p is small such that $\lambda = np$ is moderate, the Poisson distribution can be used to approximate the binomial:

$$\text{If } n \geq 20 \text{ and } p \leq 0.05, \quad \text{then } B(n,p) \approx \text{Poisson}(\lambda = np)$$

Graphical Insight

A Poisson distribution is skewed right when λ is small, but becomes more symmetric as λ increases.

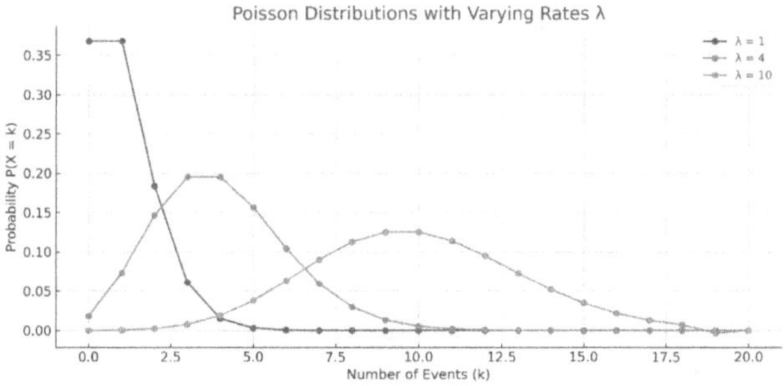

Figure 7.3: Poisson Distributions with Varying Rates λ

Summary

The Poisson distribution is a powerful tool for modeling random counts over fixed intervals. With only one parameter λ, it is simple yet versatile, and has applications across telecommunications, biology, traffic flow, reliability engineering, and operations research.

7.6 Hypergeometric Distribution

The **Hypergeometric Distribution** models the probability of obtaining a specific number of successes when sampling *without replacement* from a finite population. It is especially relevant when the population size is small and each selection affects subsequent probabilities.

7.6.1 Definition and Formula

The probability mass function (PMF) is given by:

$$P(X = x) = \frac{\binom{K}{x}\binom{N-K}{n-x}}{\binom{N}{n}}$$

where:

- N is the population size,
- K is the number of success states in the population,
- n is the number of draws,
- x is the number of observed successes,
- $\binom{a}{b}$ denotes a combination.

Conditions for Use

1. The population is finite and known.
2. Each draw is dependent (sampling without replacement).
3. There are two categories: success and failure.

7.6.2 Example 1

A factory produces 50 light bulbs, 8 of which are defective. An inspector selects 5 bulbs at random. What is the probability that exactly 2 are defective?

$$N = 50, \quad K = 8, \quad n = 5, \quad x = 2$$

$$P(X = 2) = \frac{\binom{8}{2}\binom{42}{3}}{\binom{50}{5}} = \frac{28 \cdot 11480}{2118760} \approx 0.1515$$

7.6.3 Example 2

From a deck of 52 cards, there are 13 hearts. What is the probability of drawing exactly 3 hearts in a 5-card hand (no replacement)?

$$N = 52, \quad K = 13, \quad n = 5, \quad x = 3$$

$$P(X = 3) = \frac{\binom{13}{3}\binom{39}{2}}{\binom{52}{5}} = \frac{286 \cdot 741}{2598960} \approx 0.0811$$

7.6.4 Mean and Variance

- Mean:
$$\mu = n \cdot \frac{K}{N}$$

- Variance:
$$\sigma^2 = n \cdot \frac{K}{N} \cdot \frac{N-K}{N} \cdot \frac{N-n}{N-1}$$

Feature	Binomial	Hypergeometric
Population Size	Infinite or large	Finite
Replacement	With replacement	Without replacement
Probability of Success	Constant	Varies
Independence	Independent trials	Dependent trials
Formula	$\binom{n}{x}p^x(1-p)^{n-x}$	$\frac{\binom{K}{x}\binom{N-K}{n-x}}{\binom{N}{n}}$

Table 7.1: Comparison of Binomial and Hypergeometric Distributions

Comparison with Binomial Distribution

Graphical Representation

The PMF can be visualized using a bar chart, with values of x on the x-axis and probabilities $P(X = x)$ on the y-axis. The shape of the distribution may be symmetric or skewed depending on population parameters.

Real-World Applications

- **Quality control:** Determining defective items in a batch.
- **Medical sampling:** Estimating disease prevalence in a population.
- **Ecology:** Estimating species distribution in a finite area.
- **Election auditing:** Evaluating the likelihood of ballot errors.

7.6.5 Summary

The hypergeometric distribution is essential when dealing with dependent outcomes in a finite sample space. It adjusts for the changing composition of the population after each draw, offering more accurate modeling than the binomial in such scenarios.

$$P(X = x) = \frac{\binom{K}{x}\binom{N-K}{n-x}}{\binom{N}{n}}$$

$$\mu = n \cdot \frac{K}{N}$$

$$\sigma^2 = n \cdot \frac{K}{N} \cdot \frac{N-K}{N} \cdot \frac{N-n}{N-1}$$

7.7 Multinomial Distribution

The multinomial distribution is a natural extension of the binomial distribution when an experiment has more than two possible outcomes. It describes the probabilities of various combinations of counts over multiple categories when each trial results in one of k outcomes, and the trials are independent and identically distributed.

Definition

A multinomial experiment consists of:

- A fixed number of trials, n.
- Each trial results in one of k possible outcomes (categories).
- The probability of each outcome i is denoted by p_i for $i = 1, 2, \ldots, k$, where $\sum_{i=1}^{k} p_i = 1$.
- The trials are independent.
- The random variables $X_1, X_2, \ldots, X_k$ count the number of times each outcome occurs in the n trials.

Probability Mass Function (PMF)

The probability of observing counts $x_1, x_2, \ldots, x_k$ for each of the k outcomes, where $\sum_{i=1}^{k} x_i = n$, is given by:

$$P(X_1 = x_1, \ldots, X_k = x_k) = \frac{n!}{x_1! x_2! \cdots x_k!} p_1^{x_1} p_2^{x_2} \cdots p_k^{x_k}$$

This formula computes the likelihood of one specific arrangement of outcomes over n independent trials.

Conditions

- Total number of trials: $\sum_{i=1}^{k} x_i = n$
- Total probability must be 1: $\sum_{i=1}^{k} p_i = 1$

Example

Suppose a customer can choose one of three favorite soda brands: Coke, Pepsi, or Sprite. If 10 customers are surveyed, and the probabilities of choosing each brand are:

$$p_{\text{Coke}} = 0.5, \quad p_{\text{Pepsi}} = 0.3, \quad p_{\text{Sprite}} = 0.2$$

What is the probability that exactly 5 choose Coke, 3 choose Pepsi, and 2 choose Sprite?

$$P(X_{\text{Coke}} = 5, X_{\text{Pepsi}} = 3, X_{\text{Sprite}} = 2) = \frac{10!}{5! \cdot 3! \cdot 2!} \cdot (0.5)^5 \cdot (0.3)^3 \cdot (0.2)^2$$

Expected Values and Variance

For each category i:

$$E[X_i] = np_i$$

$$\text{Var}(X_i) = np_i(1 - p_i)$$

$$\text{Cov}(X_i, X_j) = -np_ip_j \quad \text{for } i \neq j$$

Applications

Multinomial distributions are commonly used in:

- Genetics (e.g., predicting offspring combinations).

- Marketing surveys (e.g., brand preference).
- Quality control (e.g., defect categories in manufacturing).
- Natural language processing (e.g., word occurrence in text).

Interpretation

Multinomial distributions help assess the likelihood of observing a specific configuration of categorical data. When categories are equally likely and many trials are performed, it behaves similarly to a categorical extension of the normal distribution due to the Central Limit Theorem.

7.8 Conclusion

Discrete probability distributions are foundational for understanding probabilistic systems involving countable outcomes. They support decision-making in real-world scenarios from genetics to telecommunications, industrial quality control to marketing.

Exercises

1. **Binomial Basics:** A machine has a 90% chance of producing a non-defective widget. If 8 widgets are selected at random, what is the probability exactly 7 of them are non-defective?

2. **Poisson Problem:** A call center receives an average of 3 calls per minute. What is the probability that it receives exactly 5 calls in a particular minute?

3. **Hypergeometric Sampling:** A shipment of 20 computers contains 6 that are defective. If 5 computers are randomly selected without replacement, what is the probability that exactly 2 of them are defective?

4. **Multinomial Situation:** In a bag of 10 marbles, 4 are red, 3 are blue, and 3 are green. If 5 marbles are drawn with replacement, what is the probability of drawing 2 red, 2 blue, and 1 green?

5. **PMF Construction:** Let the random variable X represent the number of heads obtained when flipping a fair coin three times. Construct the probability mass function (PMF) for X.

6. **Expected Value & Variance:** Suppose a discrete random variable X has the following distribution:

$$P(X = 1) = 0.1,$$
$$P(X = 2) = 0.3,$$
$$P(X = 3) = 0.4,$$
$$P(X = 4) = 0.2$$

Compute the expected value $E(X)$ and the variance $\text{Var}(X)$.

7. **CDF Interpretation:** A random variable X has the cumulative distribution function:

$$F(x) = \begin{cases} 0, & x < 0 \\ 0.3, & 0 \leq x < 1 \\ 0.7, & 1 \leq x < 2 \\ 1, & x \geq 2 \end{cases}$$

   Find $P(0 \leq X < 2)$ and $P(X = 1)$.

8. **Mean of a Binomial:** For a binomial distribution with $n = 12$ and $p = 0.6$, find the mean μ and standard deviation σ.

9. **Poisson Approximation:** A rare event occurs on average 2 times per month. Approximate the probability that it happens 0 times in a month using the Poisson distribution.

10. **Conceptual Question:** Explain the key difference between the binomial and hypergeometric distributions in terms of sampling method and independence of trials.

Chapter 8

Continuous Variables

8.1 Introduction

A **continuous random variable** is one that can take an infinite number of values within an interval or union of intervals on the real number line. Unlike discrete random variables, where we calculate probabilities using sums and probability mass functions (PMFs), continuous random variables use integrals and probability density functions (PDFs).

One fundamental property of continuous variables is:

$$P(X = x) = 0 \quad \text{for any specific } x$$

This is because the probability is spread continuously over an interval, and no single point has positive probability.

To transition from discrete to continuous frameworks, we substitute:

- Sums $\sum$ with integrals $\int$
- PMFs $P(X = x)$ with PDFs $f(x)$

8.2 PDF

A **probability density function** (PDF) is a fundamental tool used to describe the likelihood of a continuous random variable taking on a particular value within a specified range. Unlike discrete random variables that have probability mass functions (PMFs), continuous random variables use PDFs because the probability of a continuous variable equaling any specific value is zero:

$$P(X = x) = 0 \quad \text{for all } x \in \mathbb{R}$$

Instead, we assign probabilities to intervals of values using integrals of the PDF over those intervals.

Definition

A function $f(x)$ is a valid PDF for a continuous random variable X if it satisfies the following two properties:

1. **Non-negativity:** The function must be

non-negative for all real numbers:

$$f(x) \geq 0 \quad \text{for all } x \in \mathbb{R}$$

2. **Total probability is 1:** The area under the entire curve of $f(x)$ over the real line must equal 1:

$$\int_{-\infty}^{\infty} f(x)\,dx = 1$$

Probability as Area Under the Curve

For a continuous random variable X, the probability that X falls within a certain interval $[a, b]$ is the area under the curve of the PDF between a and b. This is computed by the definite integral:

$$P(a \leq X \leq b) = \int_a^b f(x)\,dx$$

This integral represents the total probability mass in that interval. If the function $f(x)$ is flat, as in the uniform distribution, the integral becomes a simple calculation of base times height. If the function is curved, as in the normal distribution, we use calculus or statistical tables.

Example: Uniform Distribution

Let $X \sim \text{Uniform}(2, 5)$. Then:

$$f(x) = \begin{cases} \frac{1}{5-2} = \frac{1}{3} & \text{if } 2 \leq x \leq 5 \\ 0 & \text{otherwise} \end{cases}$$

To find the probability that X lies between 2.5 and 4:

$$P(2.5 \leq X \leq 4) = \int_{2.5}^{4} \frac{1}{3} dx = \frac{1}{3}(4 - 2.5) = \frac{1.5}{3} = 0.5$$

Visual Representation

You can visualize a PDF as a smooth curve above the x-axis. The total area under the curve is 1, and the area over any interval corresponds to the probability of the random variable falling in that range.

Key Notes:

- The height of the curve at a particular x does not represent the probability at that point.
- Instead, probabilities are derived from the area under the curve across an interval.
- For symmetric PDFs like the normal distribution, the mean is at the center and acts as the point of symmetry.

Piecewise PDFs

Some PDFs are defined piecewise. For example:

$$f(x) = \begin{cases} 2x & \text{for } 0 \leq x \leq 1 \\ 0 & \text{otherwise} \end{cases}$$

We verify this is a PDF by checking the total area:

$$\int_0^1 2x\,dx = \left[x^2\right]_0^1 = 1$$

Then:

$$\begin{aligned} P(0.25 \leq X \leq 0.75) &= \int_{0.25}^{0.75} 2x\,dx \\ &= \left[x^2\right]_{0.25}^{0.75} \\ &= 0.75^2 - 0.25^2 \\ &= 0.5625 - 0.0625 = 0.5 \end{aligned}$$

Relation to the CDF

The PDF is the derivative of the cumulative distribution function (CDF):

$$f(x) = \frac{d}{dx} F(x)$$

Conversely:

$$F(x) = \int_{-\infty}^{x} f(t)\,dt$$

This dual relationship is essential in practice when converting between raw densities and cumulative probabilities.

Conclusion

Understanding PDFs equips us with a powerful tool to handle real-world problems involving continuous variables — such as measuring heights, weights, or waiting times — where outcomes lie on a continuum. In subsequent sections, we will apply these principles to commonly used distributions like the exponential and normal distributions.

Example: Uniform Distribution

Let $X \sim \text{Uniform}(0, 1)$. Then:

$$f(x) = \begin{cases} 1 & \text{if } 0 \le x \le 1 \\ 0 & \text{otherwise} \end{cases}$$

$$P(0.25 \le X \le 0.75) = \int_{0.25}^{0.75} 1\,dx = 0.5$$

8.3 CDF

The **cumulative distribution function (CDF)** of a continuous random variable X, denoted $F(x)$, represents the probability that the random variable takes on a value less than or equal to x. It accumulates the probabilities from the left up to x, and is defined by:

$$F(x) = P(X \leq x) = \int_{-\infty}^{x} f(t)\, dt$$

Where $f(t)$ is the probability density function (PDF) of X.

Key Properties of the CDF

1. **Domain and Range:** The domain of $F(x)$ is $\mathbb{R}$, and the range is the interval $[0, 1]$.
2. **Monotonicity:** $F(x)$ is a non-decreasing function. That is, if $a < b$, then $F(a) \leq F(b)$.
3. **Boundary Conditions:**

   $$\lim_{x \to -\infty} F(x) = 0 \quad \text{and} \quad \lim_{x \to \infty} F(x) = 1$$

   These limits ensure that the CDF covers the full probability space.
4. **Continuity:** The CDF of a continuous random variable is always a continuous function.
5. **Relation to PDF:** If X is a continuous random

variable with PDF $f(x)$, then:

$$f(x) = \frac{d}{dx}F(x)$$

Using the CDF to Compute Probabilities

The CDF simplifies the computation of probabilities over intervals:

$$P(a \leq X \leq b) = F(b) - F(a)$$

This makes it easy to compute probabilities when a closed-form of the CDF is available.

Example 1: Uniform Distribution

Let $X \sim \text{Uniform}(2, 5)$. The PDF is:

$$f(x) = \begin{cases} \frac{1}{3}, & 2 \leq x \leq 5 \\ 0, & \text{otherwise} \end{cases}$$

Then the CDF is:

$$F(x) = \begin{cases} 0, & x < 2 \\ \frac{x-2}{3}, & 2 \leq x \leq 5 \\ 1, & x > 5 \end{cases}$$

To find $P(2.5 \leq X \leq 4.2)$, we compute:

$$\begin{aligned} F(4.2) - F(2.5) &= \frac{4.2 - 2}{3} - \frac{2.5 - 2}{3} \\ &= \frac{2.2}{3} - \frac{0.5}{3} \\ &= \frac{1.7}{3} \approx 0.5667 \end{aligned}$$

Example 2: Exponential Distribution

Let $X \sim \text{Exponential}(\lambda = 2)$. The PDF is:

$$f(x) = \begin{cases} 2e^{-2x}, & x \geq 0 \\ 0, & x < 0 \end{cases}$$

Then the CDF is:

$$F(x) = \begin{cases} 0, & x < 0 \\ 1 - e^{-2x}, & x \geq 0 \end{cases}$$

To find $P(X \leq 1)$:

$$F(1) = 1 - e^{-2(1)} = 1 - e^{-2} \approx 1 - 0.1353 = 0.8647$$

Graphical Interpretation

The CDF is plotted as a smooth, non-decreasing curve that starts at 0 and asymptotically approaches 1. At any point x, the height of the CDF curve represents the total probability accumulated from $-\infty$ up to x.

- Steeper slopes in the CDF correspond to higher density in the PDF.
- Flat regions indicate intervals where the PDF is zero.
- If the CDF has a jump, that implies a discrete component at that value.

Inverse CDF (Quantile Function)

For continuous distributions, the inverse CDF (also called the quantile function) $F^{-1}(p)$ gives the value x such that:

$$P(X \leq x) = p$$

This is useful in simulations (e.g., inverse transform sampling) and for calculating percentiles.

Conclusion

The CDF encapsulates the cumulative behavior of a random variable. It complements the PDF and provides an essential foundation for probability computations, statistical inference, and simulations. Mastering the interplay between CDFs and PDFs is crucial for working with continuous distributions.

8.4 Common Distributions

8.4.1 Uniform Distribution

The uniform distribution models a situation where every outcome in the interval $[a, b]$ is equally likely. It is the most basic form of a continuous distribution and is often used as a building block for more complex models.

$$f(x) = \begin{cases} \frac{1}{b-a}, & a \leq x \leq b \\ 0, & \text{otherwise} \end{cases}$$

$$\mu = \frac{a+b}{2}, \qquad \sigma^2 = \frac{(b-a)^2}{12}, \qquad \sigma = \frac{b-a}{\sqrt{12}}$$

Example: If the arrival time of a bus is uniformly distributed between 10:00 and 10:30 (in decimal, $a = 0$, $b = 30$ minutes), then the expected arrival time is 15 minutes, and the standard deviation is $\frac{30}{\sqrt{12}} \approx 8.66$ minutes.

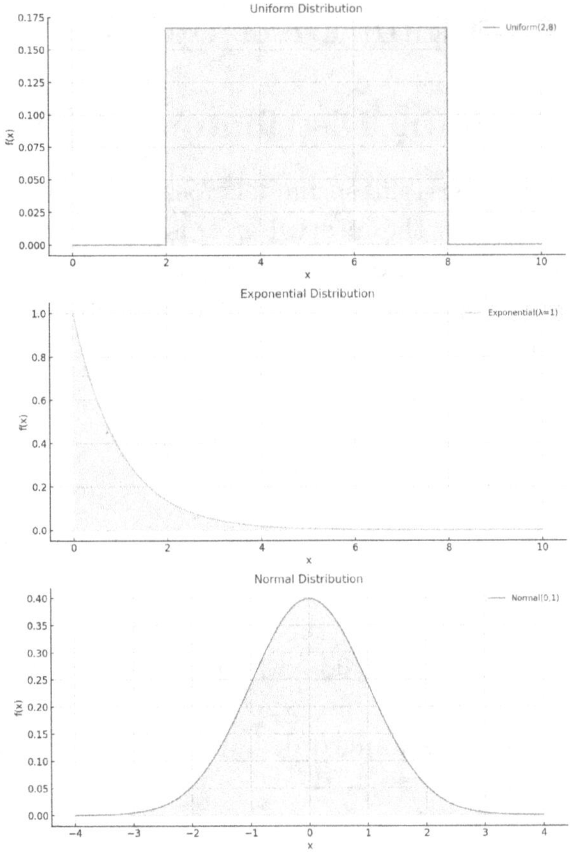

Figure 8.1: Uniform (blue), Exponential (green), and Normal (red) distribution curves.

8.4.2 Exponential Distribution

The exponential distribution is widely used to model the time between events in a Poisson process (i.e., events occurring independently at a constant average rate).

$$f(x) = \begin{cases} \lambda e^{-\lambda x}, & x \geq 0 \\ 0, & \text{otherwise} \end{cases}$$

$$\mu = \frac{1}{\lambda}, \qquad \sigma^2 = \frac{1}{\lambda^2}, \qquad \sigma = \frac{1}{\lambda}$$

Key Properties:

- **Memoryless property:**
  $P(X > s + t | X > s) = P(X > t)$
- Common in queuing theory, survival analysis, and reliability engineering

Example: Suppose lightbulbs fail at a constant average rate of $\lambda = 0.05$ failures per hour. Then the probability that a bulb lasts more than 20 hours is:

$$P(X > 20) = e^{-\lambda \cdot 20} = e^{-1} \approx 0.3679$$

8.4.3 Normal Distribution

The Gaussian distribution is a cornerstone of statistics. It arises naturally in numerous contexts due to the Central Limit Theorem. It is characterized by its symmetric, bell-shaped curve and is defined over the entire real line.

$$f(x) = \frac{1}{\sqrt{2\pi\sigma^2}} \exp\left(-\frac{(x - \mu)^2}{2\sigma^2} \right)$$

Properties:

- Symmetric about the mean μ
- Mean = median = mode
- Approximately 68% of values fall within 1σ of the mean, 95% within 2σ, and 99.7% within 3σ

Standard Normal Distribution: When $\mu = 0$ and $\sigma = 1$, the distribution is known as the **standard normal**:

$$Z \sim \mathcal{N}(0, 1)$$

Example: The height of adult women is approximately normal with $\mu = 64$ inches and $\sigma = 2.5$ inches. The probability that a randomly selected woman is taller than 67 inches is:

$$P(X > 67) = P\left(Z > \frac{67 - 64}{2.5}\right) = P(Z > 1.2) \approx 0.1151$$

Standard Normal Distribution and Z-Score Table

The **standard normal distribution** is a normal distribution with mean $\mu = 0$ and standard deviation $\sigma = 1$. It is denoted as $Z \sim N(0, 1)$. The Z-score table (or standard normal table) provides the cumulative probability from the far left up to a given Z value:

$$P(Z \leq z)$$

Z-scores allow us to standardize any normal distribution

and compare across different distributions using the transformation:

$$Z = \frac{X - \mu}{\sigma}$$

This transformation converts a raw score X into the number of standard deviations away from the mean.

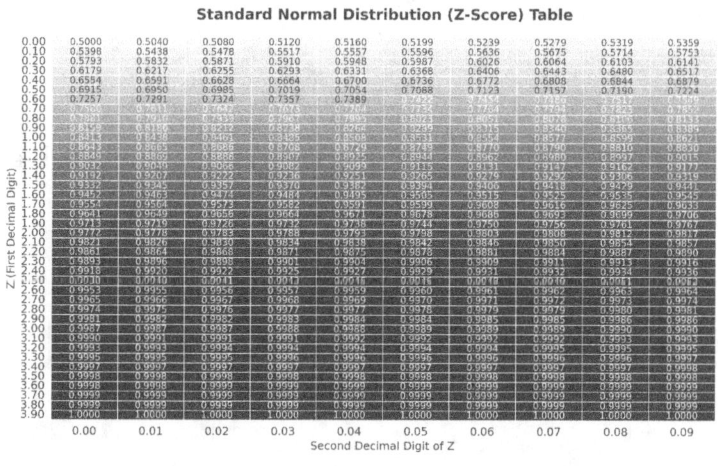

Figure 8.2: Z-Score Table: Each row shows the first two digits of the Z-score (e.g., 1.4), and each column adds the hundredths digit (e.g., 0.01). The cell at row 1.4 and column 0.01 gives $P(Z \leq 1.41)$.

Use this table to find probabilities for standard normal values. For example, to find $P(Z \leq 1.96)$, locate the row for 1.9 and the column for 0.06. The corresponding value gives the cumulative area under the standard normal curve up to $Z = 1.96$.

Visual Summary

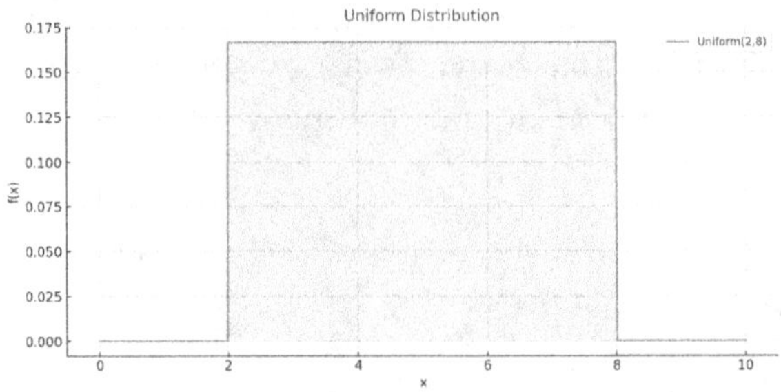

Figure 8.3: Uniform distribution: constant probability over interval $[a, b]$.

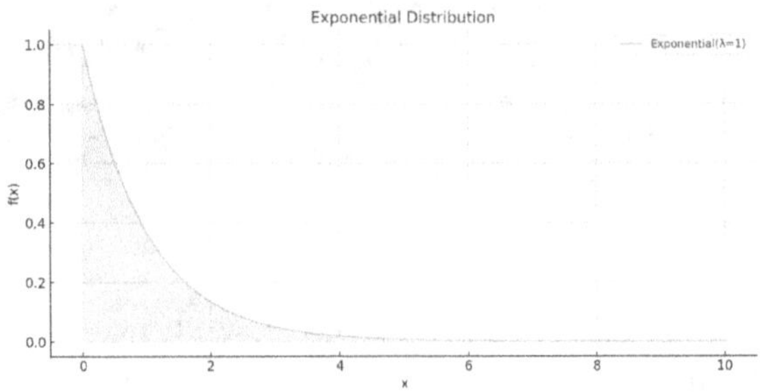

Figure 8.4: Exponential distribution: models waiting times with rapid decay.

The figure above shows:

- A **uniform distribution** (blue): constant height across a fixed interval
- An **exponential distribution** (green): rapidly decaying tail starting at $x = 0$

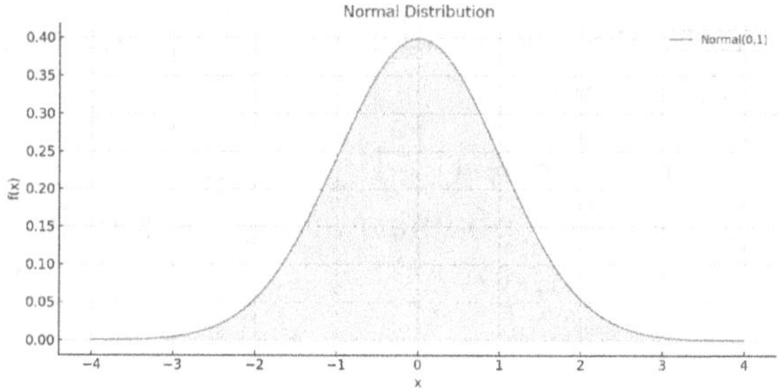

Figure 8.5: Normal distribution: bell-shaped curve centered around the mean.

- A **normal distribution** (red): symmetric bell shape centered around the mean

These foundational distributions underpin a wide variety of real-world processes and statistical inference techniques.

8.4.4 Other Common Distributions

- **Beta Distribution:** Defined on $[0, 1]$, useful for modeling proportions such as success rates in A/B testing.
- **Gamma Distribution:** A generalization of the exponential distribution, often used in queuing models, climatology, and reliability engineering.
- **Chi-Square Distribution:** A special case of the gamma distribution, primarily used in hypothesis testing and confidence intervals for variance.
- **t-Distribution:** Similar to the normal but with

heavier tails, used in small-sample inference for population means.

- **F-Distribution:** Used to compare variances across multiple populations, especially in analysis of variance (ANOVA).

8.5 Mixed Distributions

A mixed distribution contains both discrete and continuous components. For instance, consider a system that either fails immediately with probability p or has a continuous lifetime governed by an exponential distribution otherwise. The PDF of a mixed distribution is expressed as:

$$f(x) = p \cdot \delta(x) + (1 - p) \cdot f_{\text{continuous}}(x)$$

Where $\delta(x)$ represents a point mass at zero.

8.6 Summary Table

To help consolidate your understanding of the key continuous probability distributions introduced in this chapter, the table below summarizes their core properties. For each distribution, we include the probability density function (PDF), the mean (expected value), and the variance. These three characteristics are central to understanding how each distribution behaves, and how it can be applied to real-world scenarios.

Distribution	PDF $f(x)$	Mean μ	Variance σ^2
Uniform(a, b)	$\frac{1}{b-a}$	$\frac{a+b}{2}$	$\frac{(b-a)^2}{12}$
Exponential(λ)	$\lambda e^{-\lambda x}$	$\frac{1}{\lambda}$	$\frac{1}{\lambda^2}$
Normal(μ, σ^2)	see above	μ	σ^2

Uniform Distribution: This distribution assigns equal probability density across the entire interval $[a, b]$. It is often used in modeling events where all outcomes are equally likely within a range—like randomly selecting a time from a 24-hour clock.

Exponential Distribution: This models the time between independent events occurring at a constant average rate (e.g., time between bus arrivals). It is memoryless, meaning the probability of an event occurring in the future is independent of the past.

Normal Distribution: This is the famous bell-shaped curve that appears frequently in statistics. It is symmetric around the mean and described by its mean μ and variance σ^2. Most natural phenomena and measurement errors tend to follow a normal distribution, especially due to the Central Limit Theorem.

Each distribution has its own behavior and applications, and it's important to recognize when to use each. Understanding their shape, formulas, and parameters will serve as a foundation for modeling real-world uncertainty in advanced statistics and data analysis.

8.7 Exercises

The following problems are designed to help reinforce your understanding of the Uniform, Exponential, and Normal distributions. Solve each problem by clearly identifying the type of distribution, writing down the relevant formulas, and showing your steps.

1. **Uniform Distribution:** A random variable X is uniformly distributed between 10 and 30.
   (a) Find the probability that X falls between 15 and 25.
   (b) Compute the mean and standard deviation of X.
2. **Uniform Distribution:** A bus arrives randomly between 8:00 a.m. and 8:30 a.m. What is the probability that a person who arrives at 8:10 a.m. waits less than 5 minutes?
3. **Exponential Distribution:** The time between incoming calls at a call center follows an exponential distribution with a mean of 6 minutes.
   (a) What is the value of λ?
   (b) What is the probability that the next call arrives within 3 minutes?
   (c) What is the probability that the next call takes more than 10 minutes to arrive?
4. **Exponential Distribution:** Suppose the lifespan of a certain type of lightbulb is

exponentially distributed with a rate parameter $\lambda = 0.1$ hours^{-1}.

(a) Find the probability that a lightbulb lasts more than 20 hours.

(b) Find the median lifespan.

5. **Normal Distribution:** Assume that scores on a college entrance exam are normally distributed with mean $\mu = 500$ and standard deviation $\sigma = 100$.

   (a) What is the probability that a randomly selected student scores above 650?

   (b) What is the probability that a score falls between 450 and 550?

   (c) Find the score that corresponds to the 90th percentile.

6. **Normal Distribution:** The weights of a certain breed of dogs are normally distributed with mean 60 lb and standard deviation 8 lb.

   (a) What proportion of dogs weigh more than 70 lb?

   (b) What is the weight below which 25% of dogs fall?

7. **Mixed Distribution Comparison:** Briefly explain the difference in how probability is interpreted in a Uniform distribution versus an Exponential distribution. Why is $P(X = x) = 0$ in both?

8. **Application:** You are told a machine part lasts on average 1200 hours, and lifespans follow an exponential distribution.
   (a) What is the probability that a randomly selected part lasts more than 1500 hours?
   (b) What is the probability it fails within the first 500 hours?
9. **Z-Score Practice:** A student's score on a standardized math test is 680. The population mean is 600 with a standard deviation of 50.
   (a) Compute the z-score.
   (b) Using a z-table, estimate the probability that a randomly selected student scores below 680.
10. **Graphical Interpretation:** Sketch rough shapes of the following PDFs and label axes and key features:
    (a) $\text{Uniform}(0, 10)$
    (b) Exponential with $\lambda = 0.5$
    (c) $\text{Normal}(\mu = 100, \sigma = 15)$

Chapter 9

Sampling Distributions

Sampling distributions form the foundation of inferential statistics, enabling us to draw conclusions about a population based on a sample. In this chapter, we explore the sampling distribution of sample statistics, with a primary focus on the sample mean. We also cover how standard errors are used to measure sampling variability and the difference between using the normal (z) and t distributions depending on whether the population standard deviation (σ) is known. Additionally, we introduce the **chi-square distribution**, which is used when analyzing the sampling distribution of sample variances, especially in constructing confidence intervals or performing hypothesis tests about population variance.

9.1 Introduction

A sampling distribution is the probability distribution of a given statistic based on a random sample.

- When we take many samples of the same size from a population and compute a statistic (e.g., mean, proportion), the distribution of that statistic across all samples is the sampling distribution.
- The Central Limit Theorem states that, for a large enough sample size, the sampling distribution of the sample mean will be approximately normal, regardless of the population distribution.
- For statistics involving sample variance, particularly when the underlying population is normal, the sampling distribution of the sample variance follows a **chi-square distribution**. This is essential for constructing confidence intervals and performing hypothesis tests related to variance.
- For example, if a sample of size n is taken from a normal population with true variance σ^2, the statistic

  $$\chi^2 = \frac{(n-1)s^2}{\sigma^2}$$

  follows a chi-square distribution with $n - 1$ degrees of freedom. This relationship is used to test hypotheses about the population variance and to construct confidence intervals for it.

9.2 Central Limit Theorem

The **Central Limit Theorem (CLT)** is one of the most powerful results in probability and statistics. It tells us how the sampling distribution of the sample mean behaves when we draw repeated samples from a population.

Theorem 1 (Central Limit Theorem). *Let $X_1, X_2, \ldots, X_n$ be a random sample of size n drawn from a population with mean μ and finite variance σ^2. Then, as $n \to \infty$, the distribution of the standardized sample mean*

$$Z = \frac{\bar{X} - \mu}{\sigma/\sqrt{n}}$$

approaches the standard normal distribution $N(0, 1)$, regardless of the shape of the original population distribution.

Why the CLT Matters

- It allows us to make inferences about population parameters using sample data, even when the underlying population is not normally distributed.
- The approximation becomes more accurate as the sample size increases.
- For many practical purposes, $n \geq 30$ is considered sufficient for the CLT to hold.

Practical Use

Thanks to the CLT, we can construct confidence intervals and conduct hypothesis tests using the normal distribution, even if the original data is skewed or non-normal, provided the sample size is large.

9.3 Law of Large Numbers

The **Law of Large Numbers (LLN)** is a foundational principle that explains why sample averages converge to the population average.

Theorem 2 (Law of Large Numbers). *As the size of the sample n increases, the sample mean $\bar{X}$ converges in probability to the population mean μ:*

$$\lim_{n \to \infty} P(|\bar{X} - \mu| > \epsilon) = 0 \quad \textit{for any } \epsilon > 0.$$

Implications of the LLN

- The LLN justifies using the sample mean to estimate the population mean.
- Larger samples yield more reliable estimates.
- It provides the long-run stability of statistics computed from repeated trials or samples.

Comparison with CLT

While the LLN tells us that $\bar{X}$ gets closer to μ as n grows, the CLT tells us about the shape of the distribution of $\bar{X}$. They work together: the LLN ensures accuracy, and the CLT enables inference using standard distributions.

Sampling Distribution of the Sample Mean

The sampling distribution of the sample mean is one of the most fundamental concepts in statistics, particularly in inferential statistics. It describes the distribution of sample means over repeated sampling from the same population.

Let μ be the population mean and σ be the population standard deviation. Suppose $\bar{X}$ is the mean of a sample of size n drawn randomly and independently from the population.

Key Properties

- **Mean of the Sampling Distribution:**

$$E(\bar{X}) = \mu$$

  This tells us that the sample mean is an unbiased estimator of the population mean. On average, the sample mean will equal the population mean over repeated sampling.

- **Standard Deviation of the Sampling Distribution (Standard Error):**

$$\mathrm{SE}_{\bar{X}} = \frac{\sigma}{\sqrt{n}}$$

This quantity measures the variability or spread of the sample means around the true population mean. As the sample size n increases, the standard error decreases, which means our estimate becomes more precise.

Implications of the Standard Error

The standard error plays a critical role in constructing confidence intervals and conducting hypothesis tests. A smaller standard error implies greater precision in estimating the population mean. This is why larger sample sizes are desirable in statistical studies.

Effect of Sample Size

To illustrate the impact of sample size on the standard error, consider the following:

$$\mathrm{SE}_{\bar{X}} = \frac{\sigma}{\sqrt{n}} \quad \Rightarrow \quad \text{As } n \uparrow,\ \mathrm{SE}_{\bar{X}} \downarrow$$

This inverse square root relationship means that to cut the standard error in half, we must increase the sample size by a factor of four. This demonstrates diminishing returns: doubling your sample size doesn't cut your error in half — you need four times the sample size to do that.

To determine the **minimum sample size** n needed to estimate a population mean with a specified margin of error E at a given confidence level (assuming population standard deviation σ is known), we use the formula:

$$n = \left(\frac{z^* \cdot \sigma}{E}\right)^2$$

- z^* is the critical value from the standard normal distribution for the desired confidence level (e.g., $z^* = 1.96$ for 95% confidence)
- σ is the population standard deviation
- E is the desired margin of error

This formula ensures that the confidence interval will have a margin of error no greater than E, providing the sample size is at least n.

Example: To estimate a population mean within $E = 2$ units at 95% confidence with $\sigma = 10$:

$$n = \left(\frac{1.96 \cdot 10}{2}\right)^2 = \left(\frac{19.6}{2}\right)^2 = (9.8)^2 = 96.04 \Rightarrow \boxed{n = 97}$$

Sampling Distribution Examples

- If the original population is normal, then the sampling distribution of $\bar{X}$ is also normal, regardless of n.
- If the population is not normal, the Central Limit Theorem tells us that the sampling distribution of $\bar{X}$ will still be approximately normal for sufficiently large n (typically $n \geq 30$).

Practical Applications

Understanding the sampling distribution of the sample mean is essential when:

- Constructing confidence intervals for μ
- Performing hypothesis tests about the population mean
- Comparing means from different populations

Example

Suppose a population has a mean $\mu = 50$ and a standard deviation $\sigma = 10$. A sample of size $n = 25$ is drawn. The mean of the sampling distribution is:

$$E(\bar{X}) = 50$$

The standard error is:

$$\mathrm{SE}_{\bar{X}} = \frac{10}{\sqrt{25}} = 2$$

This means that, over repeated sampling, most of the sample means will fall within a range of 2 units above or below the population mean.

Visual Representation

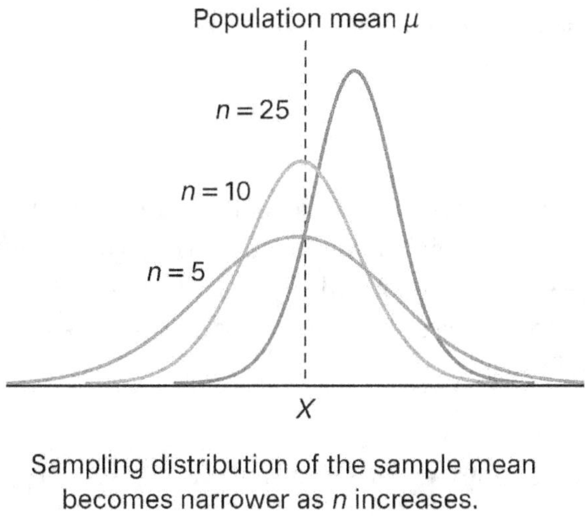

Figure 9.1: Sampling distribution of the sample mean becomes narrower as n increases.

Key Properties

- **Mean of the Sampling Distribution:**

$$E(\bar{X}) = \mu$$

  This tells us that the sample mean is an unbiased estimator of the population mean. On average, the sample mean will equal the population mean over repeated sampling.

- **Standard Deviation of the Sampling Distribution (Standard Error):**

$$\text{SE}_{\bar{X}} = \frac{\sigma}{\sqrt{n}}$$

  This quantity measures the variability or spread of the sample means around the true population mean. As the sample size n increases, the standard error decreases, which means our estimate becomes more precise.

Implications of the Standard Error

The standard error plays a critical role in constructing confidence intervals and conducting hypothesis tests. A smaller standard error implies greater precision in estimating the population mean. This is why larger sample sizes are desirable in statistical studies.

Effect of Sample Size

To illustrate the impact of sample size on the standard error, consider the following:

$$\mathrm{SE}_{\bar{X}} = \frac{\sigma}{\sqrt{n}} \quad \Rightarrow \quad \text{As } n \uparrow,\ \mathrm{SE}_{\bar{X}} \downarrow$$

This inverse square root relationship means that to cut the standard error in half, we must increase the sample size by a factor of four. This demonstrates diminishing returns: doubling your sample size doesn't cut your error in half — you need four times the sample size to do that.

Sampling Distribution Examples

- If the original population is normal, then the sampling distribution of $\bar{X}$ is also normal, regardless of n.
- If the population is not normal, the Central Limit Theorem tells us that the sampling distribution of $\bar{X}$ will still be approximately normal for sufficiently large n (typically $n \geq 30$).

Practical Applications

Understanding the sampling distribution of the sample mean is essential when:

- Constructing confidence intervals for μ

- Performing hypothesis tests about the population mean
- Comparing means from different populations

Example

Suppose a population has a mean $\mu = 50$ and a standard deviation $\sigma = 10$. A sample of size $n = 25$ is drawn. The mean of the sampling distribution is:

$$E(\bar{X}) = 50$$

The standard error is:

$$\text{SE}_{\bar{X}} = \frac{10}{\sqrt{25}} = 2$$

This means that, over repeated sampling, most of the sample means will fall within a range of 2 units above or below the population mean.

Visual Representation

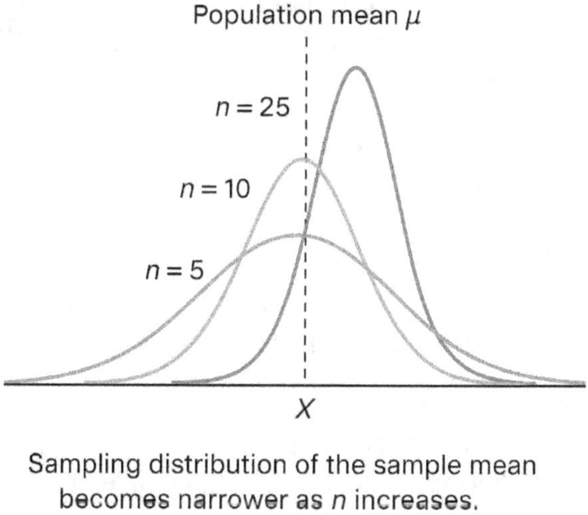

Figure 9.2: Sampling distribution of the sample mean becomes narrower as n increases.

9.4 z-Distribution

The z-distribution is one of the most fundamental tools in inferential statistics and is used extensively when working with sampling distributions, hypothesis testing, and confidence intervals. It refers to the standard normal distribution, which has a mean of 0 and a standard deviation of 1.

When Do We Use the z-Distribution?

The z-distribution is applicable when we are making inferences about a population mean, and the following conditions are met:

1. **The population standard deviation σ is known**, and
2. Either:
   - The population is **normally distributed**, regardless of sample size, or
   - The sample size is sufficiently large, typically $n \geq 30$, according to the Central Limit Theorem (CLT), which ensures that the sampling distribution of the sample mean approaches normality even if the original population distribution is not normal.

Standardizing the Sample Mean

When the above conditions are met, we can transform the sample mean $\bar{x}$ into a z-score, which tells us how many standard errors the observed sample mean is away from the hypothesized population mean μ:

$$z = \frac{\bar{x} - \mu}{\sigma/\sqrt{n}}$$

Components of the Formula

- $\bar{x}$ = observed sample mean

- μ = population mean (under the null hypothesis)
- σ = population standard deviation (known)
- n = sample size
- $\frac{\sigma}{\sqrt{n}}$ = standard error of the sample mean

The numerator $\bar{x} - \mu$ measures how far the sample mean is from the population mean. The denominator scales this distance by the standard error, allowing us to determine how "rare" or "extreme" our observation is under the assumption that the null hypothesis is true.

Visual Intuition

- The z-distribution is bell-shaped and symmetric.
- A z-score of 0 means the sample mean is exactly at the population mean.
- A z-score of $+1.96$ or -1.96 corresponds to approximately the cutoff points for a 95% confidence interval.
- The further the z-score is from 0, the more evidence we have that our sample is unusual if the null hypothesis is true.

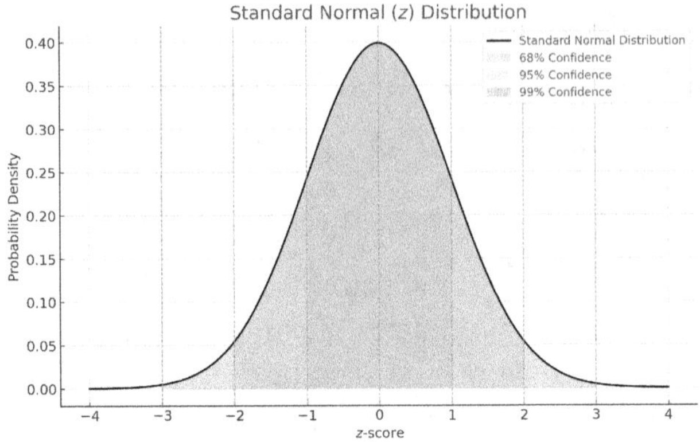

Figure 9.3: The standard normal (z) distribution. Shaded areas correspond to commonly used confidence levels (e.g., 68%, 95%, and 99%).

Application Scenarios

- Calculating confidence intervals for population means when σ is known.
- Performing one-sample z-tests to test hypotheses about population means.
- Quality control and process monitoring in manufacturing and engineering.
- Financial modeling and risk analysis.

Worked Example

Scenario: A manufacturer claims the average weight of a product is 100 grams with a known standard deviation of 4 grams. A quality control inspector takes a sample of 36 units and finds a sample mean of 98.5 grams. Is this evidence that the mean weight is different?

$$\mu = 100, \quad \sigma = 4, \quad \bar{x} = 98.5, \quad n = 36$$

$$\text{Standard Error (SE)} = \frac{4}{\sqrt{36}} = \frac{4}{6} = 0.667$$

$$z = \frac{98.5 - 100}{0.667} = \frac{-1.5}{0.667} \approx -2.25$$

A z-score of -2.25 suggests that the sample mean is 2.25 standard errors below the claimed population mean. The corresponding two-tailed p-value is approximately 0.024, which is less than 0.05, so we reject the null hypothesis at the 5% level — the evidence suggests the product may weigh less than advertised.

Important Notes

- The z-distribution assumes the population standard deviation σ is known — in real-world practice, this is rare. When σ is unknown and the sample size is small ($n < 30$), we use the t-distribution instead.
- Always check the assumptions before applying the z-distribution.

9.5 Degrees of Freedom

Degrees of freedom (df) refer to the number of independent values or quantities that can vary in a statistical calculation without violating any given constraints.

What Are Degrees of Freedom?

In most statistical formulas, we use sample data to estimate population parameters. However, when we use sample statistics like the sample mean or sample variance, we often impose constraints on the data. The degrees of freedom reflect how many data points are free to vary once certain statistics (like the sample mean) are fixed.

- For example, if you have a sample of n values and you know the sample mean, then only $(n - 1)$ of those values are free to vary. Once those are chosen, the final value is determined to maintain the calculated mean.

Why Are Degrees of Freedom Important?

Degrees of freedom are essential in statistical inference because they affect the shape of the probability distributions used in hypothesis testing and confidence intervals. The more degrees of freedom, the closer the distribution approximates the normal distribution.

Applications in Statistical Distributions

- **t-Distribution:** When estimating the population mean with an unknown population standard deviation, we use the t-distribution instead of the normal distribution. The shape of the t-distribution depends on the degrees of freedom, typically calculated as $df = n - 1$.
- **Chi-Square Distribution:** This distribution is used primarily for tests involving variances (e.g., goodness-of-fit, test for independence). The degrees of freedom often equal the number of categories minus one, or $df = n - 1$ for sample variance tests.
- **F-Distribution (to be discussed in a later chapter):** Used when comparing two variances or in analysis of variance (ANOVA). It involves two sets of degrees of freedom: one for the numerator and one for the denominator.

Visualizing the Concept

As degrees of freedom increase, the t-distribution and chi-square distribution both begin to resemble the normal distribution more closely. Fewer degrees of freedom result in thicker tails or more skewed distributions, which must be accounted for in calculations.

In conclusion, understanding degrees of freedom is crucial because they adjust for sample variability and ensure that we make accurate inferences about populations using limited sample data.

Visualizing Degrees of Freedom

Student's t-Distribution Across Degrees of Freedom

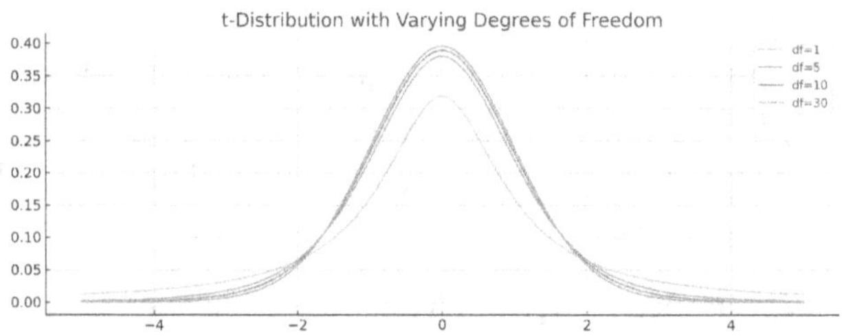

Figure 9.4: t-Distributions with varying degrees of freedom. As degrees of freedom increase, the curve approaches the standard normal distribution.

As seen above, the t-distribution is bell-shaped and symmetric, much like the standard normal distribution, but it has heavier tails. This means it accounts for more variability when sample sizes are small. As the degrees of freedom increase (e.g., larger sample sizes), the t-distribution converges to the normal distribution.

Chi-Square Distribution Across Degrees of Freedom

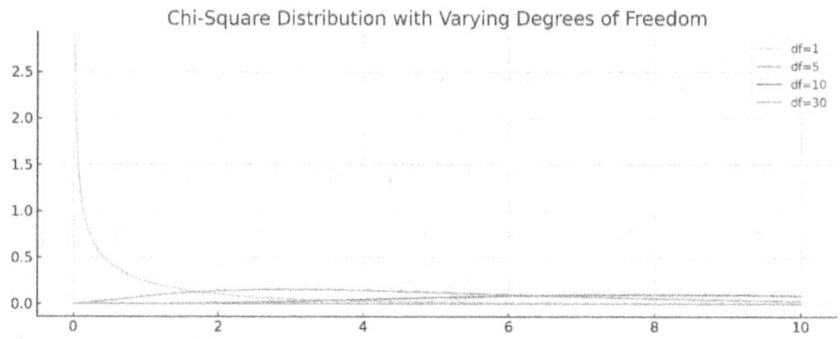

Figure 9.5: Chi-Square distributions with different degrees of freedom. The distribution becomes more symmetric and bell-shaped as degrees of freedom increase.

The Chi-Square distribution is skewed to the right and is only defined for non-negative values. It becomes more symmetric as the degrees of freedom increase. This distribution is commonly used in tests of variance, independence, and goodness of fit.

F-Distribution Across Degrees of Freedom

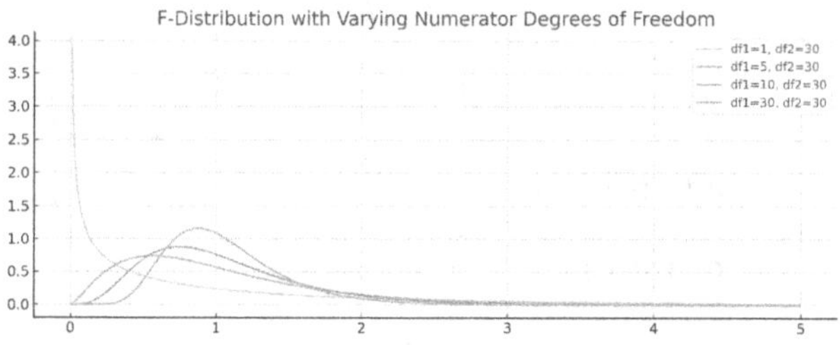

Figure 9.6: F-distributions with different numerator and denominator degrees of freedom. The shape is positively skewed, with skewness decreasing as degrees of freedom increase.

The F-distribution is used primarily in comparing variances, especially in ANOVA which we will cover in a later chapter. Its shape is influenced by two parameters: numerator degrees of freedom (d_1) and denominator degrees of freedom (d_2). As both increase, the distribution becomes less skewed and more peaked.

9.6 t-Distribution

The t-distribution is a probability distribution that arises when estimating the mean of a normally distributed population in situations where the sample size is small and the population standard deviation (σ) is unknown. Instead of using the z-distribution, which assumes a known σ, we use the t-distribution with degrees of freedom $df = n - 1$:

$$t = \frac{\bar{x} - \mu}{s/\sqrt{n}}$$

Where:

- $\bar{x}$ is the sample mean,
- μ is the population mean,
- s is the sample standard deviation,
- n is the sample size,

- $df = n - 1$ is the degrees of freedom.

The t-distribution is similar in shape to the standard normal distribution but has thicker tails. These thicker tails allow for more variability, which is appropriate for smaller sample sizes where more uncertainty exists in the estimation of σ.

As the sample size increases, the t-distribution approaches the standard normal distribution. When $n \geq 30$, the t and z distributions become nearly indistinguishable.

The t-distribution is used in a variety of hypothesis testing and confidence interval procedures, particularly:

- Confidence intervals for the mean when σ is unknown.
- One-sample t-tests for population mean.
- Two-sample t-tests comparing two means.
- Paired-sample t-tests for before-and-after measurements.

Understanding the t-distribution is crucial for applied statistics and inferential reasoning based on small samples.

Standard Errors

The **standard error (SE)** is a vital concept in inferential statistics. It quantifies the amount of variability or dispersion of a sample statistic from the true population parameter due to random sampling. When we take different samples from the same population, we expect some variation in the resulting sample statistics (e.g., means, proportions), even if there is no actual change in the population itself. This variability is what standard errors help us measure and understand.

A smaller standard error means that the sample statistic is likely to be close to the true population value, indicating more precision. Larger standard errors signal more uncertainty in the estimate and wider confidence intervals.

Standard errors differ depending on the type of statistic being calculated. Below, we break down common formulas and interpretations for the standard error of different sample statistics.

Standard Error of the Sample Mean

When estimating the population mean μ from a sample, we often use the sample mean $\bar{x}$ as an estimator. The standard error of the sample mean indicates how much variability we can expect between different sample means:

$$SE_{\bar{x}} = \frac{\sigma}{\sqrt{n}} \quad \text{(if } \sigma \text{ is known)}$$

$$SE_{\bar{x}} = \frac{s}{\sqrt{n}} \quad \text{(if } \sigma \text{ is unknown)}$$

- σ: Population standard deviation (known case).
- s: Sample standard deviation (unknown case).
- n: Sample size.

This formula highlights an important principle: as the sample size increases, the standard error decreases, following a square root relationship. This means that larger samples provide more precise estimates of the population mean.

Example: Suppose a sample of 36 students yields an average test score of 80 with a standard deviation of 12. Then:

$$SE_{\bar{x}} = \frac{12}{\sqrt{36}} = \frac{12}{6} = 2$$

This tells us that the average test score in repeated samples would typically vary by about 2 points.

Standard Error of the Sample Proportion

The standard error of a sample proportion measures how much the observed proportion in the sample ($\hat{p}$) is expected to fluctuate from the actual population proportion p:

$$SE_{\hat{p}} = \sqrt{\frac{p(1-p)}{n}}$$

- p: True population proportion (or best estimate from data).
- n: Sample size.

This formula is particularly useful for binary categorical variables (e.g., success/failure, yes/no, pass/fail).

Example: If 60 out of 100 surveyed people support a new policy, then:

$$\hat{p} = 0.60, \quad SE_{\hat{p}} = \sqrt{\frac{0.6(0.4)}{100}} = \sqrt{0.0024} \approx 0.049$$

Thus, the sample proportion has a standard error of about 4.9 percentage points.

Standard Error for Difference Between Two Means

When comparing the means of two independent groups, we calculate the standard error for the difference in sample means:

$$SE_{\bar{x}_1-\bar{x}_2} = \sqrt{\frac{\sigma_1^2}{n_1} + \frac{\sigma_2^2}{n_2}} \quad \text{(if population variances are known)}$$

$$SE_{\bar{x}_1-\bar{x}_2} = \sqrt{\frac{s_1^2}{n_1} + \frac{s_2^2}{n_2}} \quad \text{(if variances are unknown)}$$

- s_1, s_2: Sample standard deviations of the two groups.
- n_1, n_2: Sample sizes of the two groups.

This is used in two-sample z- or t-tests to evaluate

whether the means of two populations are significantly different.

Example: Suppose we compare test scores between two groups:

$$s_1 = 10,\ s_2 = 12,\ n_1 = 25,\ n_2 = 30$$

$$\begin{aligned} SE &= \sqrt{\frac{10^2}{25} + \frac{12^2}{30}} \\ &= \sqrt{\frac{100}{25} + \frac{144}{30}} \\ &= \sqrt{4 + 4.8} \\ &= \sqrt{8.8} \approx 2.97 \end{aligned}$$

Standard Error for Difference Between Two Proportions

When comparing proportions from two groups, the standard error measures the variability in the difference:

$$SE_{\hat{p}_1 - \hat{p}_2} = \sqrt{\frac{p_1(1-p_1)}{n_1} + \frac{p_2(1-p_2)}{n_2}}$$

- p_1, p_2: Sample proportions from each group.
- n_1, n_2: Corresponding sample sizes.

Example: Suppose group A has 45 successes out of 100 trials ($p_1 = 0.45$), and group B has 60 successes out of

150 trials ($p_2 = 0.40$):

$$
\begin{aligned}
SE &= \sqrt{\frac{0.45(0.55)}{100} + \frac{0.40(0.60)}{150}} \\
&= \sqrt{0.002475 + 0.0016} \\
&= \sqrt{0.004075} \\
&\approx 0.0639
\end{aligned}
$$

Why SE is Important

Understanding standard errors is foundational for:

- Constructing **confidence intervals** to estimate population parameters.
- Conducting **hypothesis tests** to evaluate statistical significance.
- Comparing variability between samples and between statistical models.
- Quantifying uncertainty in **regression analysis**, **ANOVA**, and other inferential tools.

A well-interpreted standard error helps distinguish between real effects and random noise. Always remember: **smaller SE** $\Rightarrow$ **greater precision**.

Why Standard Errors Matter

Standard errors are critical in:

- Constructing confidence intervals

- Performing hypothesis tests
- Assessing precision and uncertainty

They inform how "trustworthy" our sample estimates are and allow us to quantify the uncertainty inherent in sampling.

9.7 Unbiased Estimator

One of the most powerful results in statistics is that the sample mean, denoted as $\bar{X}$, is an **unbiased estimator** of the population mean μ. This means that if we took an infinite number of random samples from a population and computed the sample mean for each, the average of all those sample means would equal the true population mean.

$$E(\bar{X}) = \mu$$

This is crucial because it allows researchers to make inferences about population parameters even when they cannot access the full population. The concept of *expected value* extends beyond just the mean — it forms the basis for probability-weighted averages in any distribution.

For example, if the sample mean $\bar{X}$ is 52 based on a sample of size 100, and the population mean μ is actually 52, this is a direct reflection of the unbiased nature of $\bar{X}$.

Key properties of expected value:

- Linearity: $E(aX + b) = aE(X) + b$
- If X and Y are random variables, then $E(X + Y) = E(X) + E(Y)$

In practical terms, this means we can trust sample means — over time and across studies — to target the correct population value, especially when we use proper sampling methods.

Visualizing z and t Distributions

In inferential statistics, two critical distributions emerge when estimating population parameters from samples: the **standard normal distribution** (also called the z-distribution) and the **Student's t-distribution**. While both are bell-shaped and symmetric around zero, they serve different roles depending on our knowledge of the population.

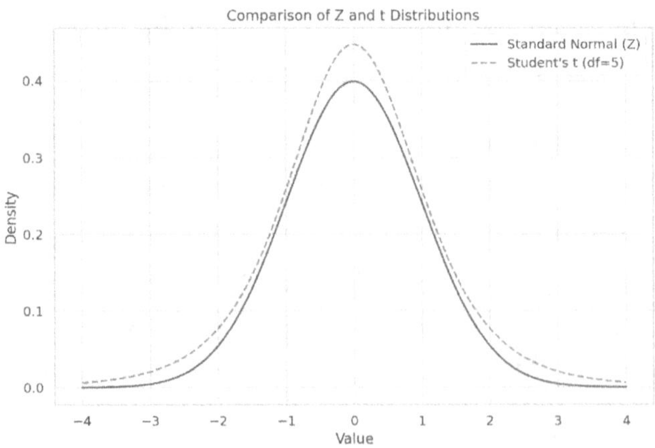

Figure 9.7: Comparison of the z and t distributions: The z-distribution (blue line) has less variability than the t-distribution (green dashed line) when degrees of freedom are low (e.g., df = 5).

Standard Normal (z) Distribution

We use the z-distribution when:

- The population standard deviation σ is known.
- The population is normally distributed or the sample size is large ($n \geq 30$).

$$z = \frac{\bar{x} - \mu}{\sigma/\sqrt{n}}$$

This gives us a standardized score that tells us how many standard errors away our sample mean $\bar{x}$ is from the population mean μ.

Student's t-Distribution

We use the t-distribution when:

- The population standard deviation σ is unknown.
- The sample size is small ($n < 30$).

The t-distribution is similar in shape to the z-distribution but has **heavier tails**. This means it accounts for more uncertainty in smaller samples. As the degrees of freedom increase (which generally corresponds to increasing sample size), the t-distribution approaches the standard normal.

$$t = \frac{\bar{x} - \mu}{s/\sqrt{n}} \quad \text{where } s \text{ is the sample standard deviation}$$

Key distinctions:

- The t-distribution adjusts for increased variability in small samples.
- It depends on degrees of freedom ($df = n - 1$).
- It becomes practically indistinguishable from the z-distribution as $n \to \infty$.

This distinction is crucial when deciding which test statistic to use for confidence intervals or hypothesis tests involving means.

Using the Chi-Square Distribution

The chi-square (χ^2) distribution is used in statistics primarily for hypothesis testing, particularly in tests of independence and goodness-of-fit. It is also used to construct confidence intervals for a population variance when the population is normally distributed.

A chi-square distribution with k degrees of freedom has the following properties:

- It is a continuous, right-skewed distribution.

- As the degrees of freedom increase, the distribution becomes more symmetric and approaches a normal distribution.
- The mean is equal to the degrees of freedom: $\mu = k$.
- The variance is $2k$.

The probability density function (PDF) of the chi-square distribution is:

$$f(x;k) = \frac{1}{2^{k/2}\Gamma(k/2)} x^{(k/2-1)} e^{-x/2}, \quad x > 0$$

where k is the number of degrees of freedom and Γ is the gamma function.

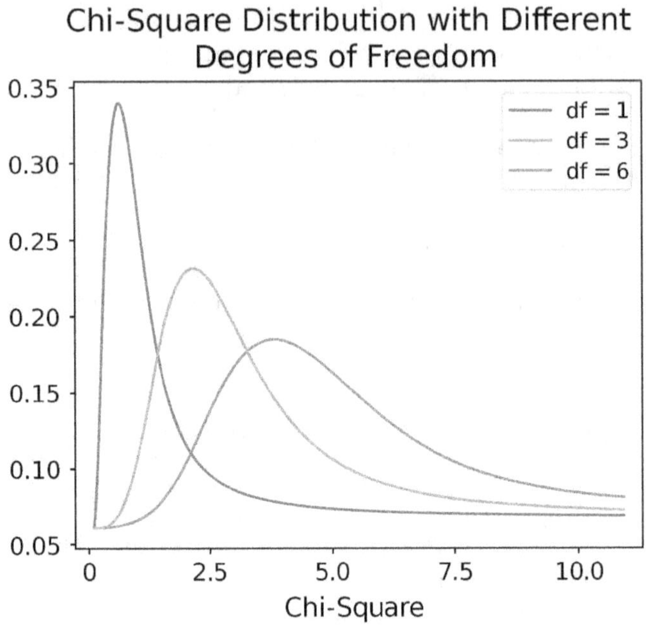

Figure 9.8: Example of chi-square distribution curves for different degrees of freedom. As k increases, the distribution becomes more symmetric.

Applications

- **Goodness-of-Fit Test:** Determines how well observed data fit an expected distribution.
- **Test of Independence:** Evaluates whether two categorical variables are independent.
- **Confidence Intervals for Variance:** For normal distributions, confidence intervals for

population variance are based on the chi-square distribution.

9.8 Summary

- The **sampling distribution** is the probability distribution of a sample statistic computed from repeated samples.
- The **mean** of the sampling distribution of the sample mean ($\bar{x}$) is equal to the population mean (μ): $E(\bar{X}) = \mu$.
- The **standard deviation** of the sampling distribution is called the **standard error**, and it varies by statistic:
  - Sample mean: $SE_{\bar{x}} = \frac{\sigma}{\sqrt{n}}$ or $\frac{s}{\sqrt{n}}$ if σ is unknown.
  - Sample proportion: $SE_{\hat{p}} = \sqrt{\frac{p(1-p)}{n}}$
  - Difference of means: $SE_{\bar{x}_1 - \bar{x}_2} = \sqrt{\frac{\sigma_1^2}{n_1} + \frac{\sigma_2^2}{n_2}}$
  - Difference of proportions: $SE_{\hat{p}_1 - \hat{p}_2} = \sqrt{\frac{p_1(1-p_1)}{n_1} + \frac{p_2(1-p_2)}{n_2}}$
- Use the z**-distribution** if σ is known and the population is normal or $n \geq 30$.
- Use the t**-distribution** if σ is unknown, particularly with small sample sizes or unknown population variance.
- The **degrees of freedom** affect the shape of the t

and chi-square distributions:

  - For t-distributions: $df = n - 1$
  - For chi-square: df equals the number of independent pieces of information used to estimate a parameter.

- The **Central Limit Theorem** (CLT) states that the sampling distribution of the sample mean approaches a normal distribution as the sample size increases, regardless of the population's shape.
- The **Law of Large Numbers** ensures that as sample size increases, the sample mean tends to converge to the population mean.
- The **chi-square distribution** is used in categorical data analysis and goodness-of-fit tests. It is right-skewed and determined by its degrees of freedom.
- As the sample size increases, the t-distribution becomes more like the z-distribution and standard errors become smaller.
- Minimum required sample sizes can be estimated using algebraic formulas to meet specific precision or confidence targets.

Exercises

1. A population has a mean of $\mu = 120$ and a standard deviation of $\sigma = 15$. What is the standard error of the sample mean for a sample

size of $n = 25$?

2. The average number of hours college students sleep is 6.8 hours with a standard deviation of 1.2 hours. A random sample of 49 students is selected. What is the probability that the sample mean is greater than 7 hours?

3. A population is skewed right. A researcher takes a sample of size $n = 40$. Explain why the sampling distribution of the sample mean is approximately normal.

4. The proportion of voters who support a new policy is $p = 0.48$. What is the standard error of the sample proportion for a sample of size $n = 150$?

5. A study measures the cholesterol levels of two different groups. Group A: $n_1 = 60$, $\bar{x}_1 = 210$, $s_1 = 25$. Group B: $n_2 = 55$, $\bar{x}_2 = 195$, $s_2 = 30$. Compute the standard error for the difference between the two sample means.

6. Suppose a sample of $n = 16$ observations is drawn from a normal population with unknown σ. The sample mean is $\bar{x} = 54.3$ and sample standard deviation is $s = 6.2$. Calculate the t-statistic for testing $H_0 : \mu = 50$.

7. A chi-square statistic of 10.21 was calculated from a sample with 5 degrees of freedom. Use a chi-square distribution table to determine if this value is significant at the $\alpha = 0.05$ level.

8. Define the Central Limit Theorem. Why is it

important when conducting hypothesis tests and constructing confidence intervals?

9. A marketing analyst wants to be 95% confident that the sample mean is within 2 units of the true mean. The population standard deviation is known to be $\sigma = 10$. What minimum sample size is needed?

10. Two independent samples are drawn to compare proportions. In sample 1, 38 out of 80 prefer option A. In sample 2, 52 out of 100 prefer option A. Compute the standard error for the difference in proportions.

Chapter 10

Confidence Intervals

Introduction

Estimation is a core task of inferential statistics: using data from a sample to draw conclusions about a population parameter. Confidence intervals (CIs) provide a range of plausible values for unknown population parameters (like a mean or proportion) and express the degree of uncertainty inherent in any estimate.

- A **point estimate** is a single number that estimates a population parameter (e.g., $\bar{x}$ for μ or $\hat{p}$ for p).
- A **confidence interval** provides a range, calculated from sample data, that is likely to contain the true population parameter.

- The **confidence level** (e.g., 90%, 95%, 99%) indicates how confident we are that the interval captures the true parameter in repeated sampling.

Example: A 95% confidence interval for a population mean might be (72.3, 78.9), meaning we are 95% confident that the true population mean lies within this range.

This chapter builds on the concepts of sampling distributions and standard errors introduced previously. We will learn:

- How to construct confidence intervals for population means and proportions.
- When to use z or t intervals.
- How to interpret confidence intervals correctly.
- How to compute the necessary sample size for a desired margin of error.

Common Point Estimates

Point estimates are single-number summaries of sample data used to estimate population parameters. Here are the most common point estimates and what they represent:

- **Sample Mean ($\bar{x}$):** Estimates the population mean (μ).
- **Sample Proportion ($\hat{p}$):** Estimates the population proportion (p).

- **Sample Variance (s^2):** Estimates the population variance (σ^2).
- **Sample Standard Deviation (s):** Estimates the population standard deviation (σ).
- **Difference of Sample Means ($\bar{x}_1 - \bar{x}_2$):** Estimates the difference between two population means ($\mu_1 - \mu_2$).
- **Difference of Sample Proportions ($\hat{p}_1 - \hat{p}_2$):** Estimates the difference between two population proportions ($p_1 - p_2$).
- **Sample Correlation Coefficient (r):** Estimates the population correlation coefficient (ρ).
- **Regression Coefficient Estimates ($\hat{\beta}_0$, $\hat{\beta}_1$, etc.):** Estimate the coefficients in a population regression model.

Each point estimate is calculated from a sample and serves as a best guess for the corresponding population parameter. These estimates form the foundation for building confidence intervals and conducting hypothesis tests.

CI for a Population Mean (σ Known)

When the population standard deviation σ is known, and the population is normally distributed or the sample size is sufficiently large ($n \geq 30$), we can use the z-distribution to construct a confidence interval for the population mean μ. The formula is:

$$\bar{x} \pm z^* \left(\frac{\sigma}{\sqrt{n}} \right)$$

Where:

- $\bar{x}$ is the sample mean,
- σ is the known population standard deviation,
- n is the sample size,
- z^* is the critical value from the standard normal distribution corresponding to the desired level of confidence (e.g., $z^* = 1.96$ for a 95% CI).

This interval provides a range of plausible values for the population mean. The interpretation is that, in repeated sampling, a certain percentage (e.g., 95%) of such intervals would capture the true population mean. This method is powerful but less commonly used in practice since σ is rarely known. It is often used in teaching to introduce the concept of interval estimation.

CI for a Population Mean (σ Unknown)

In practice, the population standard deviation σ is usually unknown. In such cases, we estimate σ using the sample standard deviation s and use the

t-distribution with degrees of freedom $df = n - 1$ to construct the confidence interval:

$$\bar{x} \pm t^* \left(\frac{s}{\sqrt{n}} \right)$$

Where:

- $\bar{x}$ is the sample mean,
- s is the sample standard deviation,
- n is the sample size,
- t^* is the critical value from the t-distribution for $df = n - 1$ at the desired confidence level.

The t-distribution is wider and has heavier tails than the z-distribution, especially for small n. As the sample size increases, the t-distribution approaches the normal distribution. This method is used in most real-world applications and forms the basis for parametric inference about the mean.

CI for a Population Proportion

When estimating a population proportion p using a sample proportion $\hat{p}$, we can construct a confidence interval using the formula:

$$\hat{p} \pm z^* \sqrt{\frac{\hat{p}(1 - \hat{p})}{n}}$$

Where:

- $\hat{p}$ is the sample proportion,
- n is the sample size,
- z^* is the critical value from the standard normal distribution.

Conditions for use:

- $n\hat{p} \geq 10$
- $n(1-\hat{p}) \geq 10$

These are known as the success-failure conditions and ensure that the sampling distribution of $\hat{p}$ is approximately normal. This method is widely used in surveys, polls, and experiments to estimate the proportion of individuals exhibiting a certain trait or behavior within a population.

Choosing the Sample Size

To determine the minimum sample size n needed for a given margin of error E and confidence level:

- For means (when σ known):

$$n = \left(\frac{z^*\sigma}{E}\right)^2$$

- For proportions:

$$n = \left(\frac{z^*}{E}\right)^2 \hat{p}(1 - \hat{p})$$

  Use $\hat{p} = 0.5$ if no prior estimate is available, to ensure a conservative (larger) sample size.

Interpreting Confidence Intervals

A 95% confidence interval means that, in the long run, 95% of such intervals constructed from random samples would contain the true population parameter. It does *not* mean there's a 95% probability that the parameter lies within a specific interval calculated from one sample.

Common Misinterpretations

- Thinking the confidence level refers to the probability the parameter is inside *this* specific interval.
- Believing a higher confidence level always produces better precision (it actually results in wider intervals).
- Confusing a lack of overlap in intervals with a definitive difference between parameters.

Exercises

1. A researcher collects a random sample of 64 college students and finds the average study time per week is 14.5 hours with a known population standard deviation of 3.2 hours. Construct a 95% confidence interval for the true mean study time of all college students.

2. In a survey of 500 voters, 280 said they support a new education bill. Construct a 99% confidence interval for the true proportion of voters who support the bill. Check whether the conditions for inference are satisfied.

3. A nutritionist wants to estimate the average sodium content in a type of canned soup. A random sample of 40 cans has a mean sodium content of 890 mg and a sample standard deviation of 120 mg. Construct a 90% confidence interval for the population mean sodium content.

4. A quality control analyst compares two production lines. Line A has 25 items with a mean weight of 6.1g and standard deviation 0.4g. Line B has 30 items with a mean weight of 6.3g and standard deviation 0.5g. Construct a 95% confidence interval for the difference in mean weights between Line A and Line B.

5. A sociologist claims that at least 60% of residents in a city support a new housing policy. A random sample of 200 residents finds that 108 support the policy. Use the sample data to construct a 95%

confidence interval for the population proportion. Does this interval support the sociologist's claim?

Chapter 11

Tests of Hypothesis

Introduction

Hypothesis testing is a formal procedure in statistics used to assess evidence provided by data. At its core, hypothesis testing enables researchers to make inferences about populations using sample data. This process allows us to determine whether there is enough statistical evidence in favor of a particular belief (called the alternative hypothesis) as opposed to a default assumption (called the null hypothesis).

We begin with a claim or assumption about a population parameter and use data from a sample to test that claim. The decision-making process involves comparing the observed data to what we would expect to observe if the null hypothesis were true. If the discrepancy is sufficiently large, we reject the null hypothesis in favor of the alternative.

Hypothesis tests are used in many fields, including medicine (e.g., testing the efficacy of a new drug), business (e.g., testing customer behavior), engineering (e.g., quality control), and social sciences.

Key Terms

- **Null Hypothesis (H_0)**: The default statement that there is no effect or no difference.
- **Alternative Hypothesis (H_1 or H_a)**: The statement that we are trying to find evidence for.
- **Test Statistic**: A standardized value calculated from sample data, used to decide whether to reject H_0.
- **P-value**: The probability of observing a test statistic as extreme as the one observed, under the assumption that H_0 is true.
- **Significance Level (α)**: A threshold for determining whether the p-value is small enough to reject H_0.
- **Type I Error**: Rejecting H_0 when it is actually true (false positive).
- **Type II Error**: Failing to reject H_0 when it is actually false (false negative).

General Procedure for Hypothesis Testing

1. State the null and alternative hypotheses.
2. Choose a significance level α (common values: 0.05, 0.01).
3. Collect data and calculate the test statistic.
4. Find the p-value or critical value.
5. Compare the p-value to α, or test statistic to critical value.
6. Make a decision: reject or fail to reject H_0.
7. Interpret the result in the context of the original problem.

This systematic framework provides a consistent method to evaluate questions about means, proportions, variances, and more. In the sections that follow, we apply this framework to one-sample and two-sample tests involving both means and proportions.

One-Sample z-Test for a Mean

The one-sample z-test for a mean is used when the population standard deviation σ is known, and either the population is normally distributed or the sample size is large ($n \geq 30$). This test evaluates whether the mean of a sample differs significantly from a known or hypothesized population mean.

Purpose

This test helps us determine if a sample mean provides enough evidence to conclude that the true population mean is different from some hypothesized value μ_0. It is often used in industrial quality control, scientific experimentation, and public health studies when population variability is known.

Assumptions

- The population standard deviation σ is known.
- The sample is a simple random sample.
- Either the population is normally distributed, or the sample size is large enough for the Central Limit Theorem to apply.

Test Statistic

The test statistic is calculated as:

$$z = \frac{\bar{x} - \mu_0}{\sigma/\sqrt{n}}$$

Where:

- $\bar{x}$ is the sample mean
- μ_0 is the hypothesized population mean
- σ is the known population standard deviation
- n is the sample size

Hypotheses

- **Two-tailed test:** $H_0 : \mu = \mu_0$ vs. $H_a : \mu \neq \mu_0$
- **Left-tailed test:** $H_0 : \mu = \mu_0$ vs. $H_a : \mu < \mu_0$
- **Right-tailed test:** $H_0 : \mu = \mu_0$ vs. $H_a : \mu > \mu_0$

Decision Rule

Using the significance level α, determine the critical z-value(s) corresponding to the type of test:

- Two-tailed: reject H_0 if $|z| > z_{\alpha/2}$
- Left-tailed: reject H_0 if $z < -z_\alpha$
- Right-tailed: reject H_0 if $z > z_\alpha$

P-value Approach

Calculate the z-statistic and find the associated p-value from the standard normal distribution. Compare p to the significance level α:

- If $p < \alpha$, reject H_0
- If $p \geq \alpha$, do not reject H_0

Confidence Interval Approach

A $(1 - \alpha)100\%$ confidence interval for μ is:

$$\bar{x} \pm z^* \left(\frac{\sigma}{\sqrt{n}} \right)$$

If μ_0 lies outside the interval, we reject H_0.

Worked Example

Suppose a machine is designed to fill bottles with 500 mL of liquid. Historically, the standard deviation is known to be $\sigma = 5$ mL. A quality control manager takes a random sample of $n = 36$ bottles and finds a sample mean of $\bar{x} = 498.7$ mL. Test at the $\alpha = 0.05$ level whether the machine is underfilling.

- $H_0 : \mu = 500,\ H_a : \mu < 500$
- $z = \frac{498.7-500}{5/\sqrt{36}} = \frac{-1.3}{0.8333} \approx -1.56$
- Critical value: $z_{0.05} = -1.645$
- Since $-1.56 > -1.645$, we fail to reject H_0

Conclusion: There is not enough evidence at the 5% level to conclude the machine is underfilling.

Common Mistakes to Avoid

- Using the z-test when σ is unknown. If σ is unknown and must be estimated from the sample, use a t-test instead.
- Misinterpreting a p-value as the probability that H_0 is true.
- Forgetting to check assumptions such as normality or adequate sample size.

One-Sample t-Test for a Mean

The one-sample t-test for a mean is used when the population standard deviation σ is unknown and must be estimated from the sample data. This test is extremely useful in real-world scenarios where population parameters are rarely known. It allows us to infer whether the sample mean is significantly different from a hypothesized population mean, even with small sample sizes.

Purpose

This test helps evaluate whether there is statistically significant evidence that the population mean differs from a known or assumed value, based on sample information and acknowledging uncertainty due to using the sample standard deviation s.

Assumptions

- The sample is a simple random sample.
- The population is approximately normally distributed, or the sample size is moderate to large (usually $n \geq 30$ for robustness).
- The population standard deviation σ is unknown.

Test Statistic

$$t = \frac{\bar{x} - \mu_0}{s/\sqrt{n}}$$

Where:

- $\bar{x}$ is the sample mean
- μ_0 is the hypothesized population mean
- s is the sample standard deviation
- n is the sample size

The degrees of freedom (df) for the t-distribution is $df = n - 1$.

Hypotheses

- **Two-tailed:** $H_0 : \mu = \mu_0$ vs. $H_a : \mu \neq \mu_0$
- **Left-tailed:** $H_0 : \mu = \mu_0$ vs. $H_a : \mu < \mu_0$
- **Right-tailed:** $H_0 : \mu = \mu_0$ vs. $H_a : \mu > \mu_0$

Decision Rule

- Determine t^* from the t-distribution table with $df = n - 1$.
- Compare the calculated t-statistic to the critical value(s):
  - Two-tailed: reject H_0 if $|t| > t_{\alpha/2,df}$
  - Left-tailed: reject H_0 if $t < -t_{\alpha,df}$

- Right-tailed: reject H_0 if $t > t_{\alpha,df}$

P-value Approach

- Use a calculator or software to find the p-value for the t-statistic.
- Compare p to α:
    - If $p < \alpha$, reject H_0
    - If $p \geq \alpha$, fail to reject H_0

Confidence Interval Approach

The $(1 - \alpha)100\%$ confidence interval for the population mean is:

$$\bar{x} \pm t^* \left(\frac{s}{\sqrt{n}}\right)$$

If μ_0 falls outside this interval, H_0 is rejected.

Worked Example

A professor believes students spend an average of 12 hours per week studying. A random sample of 15 students reported a mean of $\bar{x} = 13.2$ hours with a standard deviation of $s = 2.5$ hours. Test at the 0.05 significance level whether students study more than 12 hours.

Step 1: Hypotheses

- $H_0 : \mu = 12$

- $H_a : \mu > 12$

Step 2: Test Statistic

$$t = \frac{13.2 - 12}{2.5/\sqrt{15}} = \frac{1.2}{0.6455} \approx 1.86$$

Step 3: Critical Value With $df = 14$, $t_{0.05,14} \approx 1.761$

Step 4: Decision

$$1.86 > 1.761 \Rightarrow \text{Reject } H_0$$

Conclusion: There is sufficient evidence at the 5% level to support the claim that students study more than 12 hours per week.

Common Mistakes to Avoid

- Confusing the t-test and z-test when σ is unknown.
- Failing to use the correct degrees of freedom.
- Applying the test to small samples without checking for normality.
- Using a one-tailed test when a two-tailed test is warranted.

One-Sample z-Test for a Proportion

The one-sample z-test for a proportion is a statistical method used to test hypotheses about a single population proportion. It helps determine whether a sample proportion is significantly different from a claimed or known population proportion.

Purpose

This test is useful when you want to evaluate claims about categorical variables, such as the proportion of people who prefer a particular product, vote a certain way, or support a specific policy.

Assumptions

- The data is from a simple random sample.
- The sampling distribution of the sample proportion is approximately normal. This is satisfied when both $np \geq 10$ and $n(1-p) \geq 10$.
- The sample size is sufficiently large to apply the Central Limit Theorem to proportions.

Test Statistic

$$z = \frac{\hat{p} - p_0}{\sqrt{\frac{p_0(1-p_0)}{n}}}$$

Where:

- $\hat{p}$ is the sample proportion
- p_0 is the hypothesized population proportion
- n is the sample size

Hypotheses

- **Two-tailed:** $H_0 : p = p_0$ vs. $H_a : p \neq p_0$
- **Left-tailed:** $H_0 : p = p_0$ vs. $H_a : p < p_0$
- **Right-tailed:** $H_0 : p = p_0$ vs. $H_a : p > p_0$

Decision Rule

- Determine z^* from the standard normal distribution corresponding to the significance level α.
- Compare the calculated z-statistic to the critical value(s):
  - Two-tailed: reject H_0 if $|z| > z_{\alpha/2}$
  - Left-tailed: reject H_0 if $z < -z_\alpha$
  - Right-tailed: reject H_0 if $z > z_\alpha$

P-value Approach

- Use a calculator or statistical software to compute the p-value.
- Compare the p-value to α:

- If $p < \alpha$, reject H_0
- If $p \geq \alpha$, fail to reject H_0

Worked Example

A cell phone company claims that 75% of its users are satisfied with its service. A random sample of 120 users shows that 84 are satisfied. Is there enough evidence to doubt the company's claim at the 5% significance level?

Step 1: Hypotheses

- $H_0 : p = 0.75$
- $H_a : p \neq 0.75$

Step 2: Check Conditions

$$np_0 = 120 \times 0.75 = 90 \geq 10, \quad n(1-p_0) = 120 \times 0.25 = 30 \geq 10$$

Conditions met.

Step 3: Test Statistic

$$\hat{p} = \frac{84}{120} = 0.70$$

$$
\begin{aligned}
z &= \frac{0.70 - 0.75}{\sqrt{0.75 \times 0.25/120}} \\
&= \frac{-0.05}{\sqrt{0.1875/120}} \\
&= \frac{-0.05}{\sqrt{0.0015625}} \\
&= \frac{-0.05}{0.0395} \\
&\approx -1.27
\end{aligned}
$$

Step 4: Critical Value At $\alpha = 0.05$, $z_{\alpha/2} = 1.96$

Step 5: Decision

$$|-1.27| < 1.96 \Rightarrow \text{Fail to reject } H_0$$

Conclusion: There is not enough evidence at the 5% significance level to conclude that the satisfaction rate is different from 75%.

Confidence Interval Approach

A 95% confidence interval for p is:

$$
\begin{aligned}
\hat{p} \pm z^* \sqrt{\frac{\hat{p}(1-\hat{p})}{n}} &= 0.70 \pm 1.96 \sqrt{\frac{0.70 \times 0.30}{120}} \\
&= 0.70 \pm 1.96 \times \sqrt{0.0021} \\
&= 0.70 \pm 1.96 \times 0.0458 \\
&= 0.70 \pm 0.0897
\end{aligned}
$$

$$\text{Interval: } [0.6103,\ 0.7897]$$

Since 0.75 is within the interval, we do not reject H_0.

Common Pitfalls

- Substituting $\hat{p}$ into the standard error formula instead of using p_0 when computing the z-statistic.
- Ignoring the check for normality conditions (np, $n(1-p)$).
- Misinterpreting the p-value as the probability that H_0 is true.

Two-Sample t-Test for Means (Independent Samples)

The two-sample t-test for means is used to compare the means from two independent groups to determine whether there is a statistically significant difference between them. This is especially useful in experimental and observational studies involving different populations.

Purpose

This test helps answer questions such as: Do students from School A score higher on average than students from School B? Is the mean blood pressure different between patients on two different medications?

Assumptions

- The two samples are independent (no pairing).
- The data in each group are approximately normally distributed.
- The populations have equal or unequal variances, depending on the version of the test.
- The data are from simple random samples or randomized experiments.

Hypotheses

- **Two-tailed:** $H_0 : \mu_1 = \mu_2$ vs. $H_a : \mu_1 \neq \mu_2$
- **Left-tailed:** $H_0 : \mu_1 = \mu_2$ vs. $H_a : \mu_1 < \mu_2$
- **Right-tailed:** $H_0 : \mu_1 = \mu_2$ vs. $H_a : \mu_1 > \mu_2$

Test Statistic (Unequal Variances – Welch's t-test)

$$t = \frac{\bar{x}_1 - \bar{x}_2}{\sqrt{\frac{s_1^2}{n_1} + \frac{s_2^2}{n_2}}}$$

Where:

- $\bar{x}_1$, $\bar{x}_2$ are the sample means
- s_1, s_2 are the sample standard deviations
- n_1, n_2 are the sample sizes

Degrees of Freedom (Welch-Satterthwaite Approximation)

$$df = \frac{\left(\frac{s_1^2}{n_1} + \frac{s_2^2}{n_2}\right)^2}{\frac{\left(\frac{s_1^2}{n_1}\right)^2}{n_1-1} + \frac{\left(\frac{s_2^2}{n_2}\right)^2}{n_2-1}}$$

Note: Always round down the degrees of freedom when using a table.

Decision Rule

- Find the critical value t^* from the t-distribution for the appropriate degrees of freedom and significance level α.
- Compare the calculated t-statistic to the critical value.
- Alternatively, compute the p-value and compare it to α.

Worked Example

Researchers want to know if two different study techniques lead to different average test scores. A sample of 25 students using Technique A had a mean score of 78 with a standard deviation of 10. A sample of 20 students using Technique B had a mean score of 72 with a standard deviation of 12. Test the hypothesis at the 5% significance level.

Step 1: Hypotheses

$$H_0 : \mu_1 = \mu_2, \quad H_a : \mu_1 \neq \mu_2$$

Step 2: Test Statistic

$$\begin{aligned} t &= \frac{78 - 72}{\sqrt{\frac{10^2}{25} + \frac{12^2}{20}}} \\ &= \frac{6}{\sqrt{\frac{100}{25} + \frac{144}{20}}} \\ &= \frac{6}{\sqrt{4 + 7.2}} \\ &= \frac{6}{\sqrt{11.2}} \\ &= \frac{6}{3.3466} \\ &\approx 1.793 \end{aligned}$$

Step 3: Degrees of Freedom

$$\begin{aligned} df &= \frac{(4 + 7.2)^2}{\frac{4^2}{24} + \frac{7.2^2}{19}} \\ &= \frac{11.2^2}{\frac{16}{24} + \frac{51.84}{19}} \\ &= \frac{125.44}{0.6667 + 2.7284} \\ &= \frac{125.44}{3.3951} \\ &\approx 36.95 \Rightarrow df = 36 \end{aligned}$$

Step 4: Critical Value and Conclusion Using $df = 36$ and $\alpha = 0.05$, the two-tailed $t^* \approx 2.028$

$$|1.793| < 2.028 \Rightarrow \text{Fail to reject } H_0$$

Conclusion: There is no statistically significant difference in the mean test scores between the two techniques at the 5% level.

Confidence Interval

We can construct a 95% confidence interval for the difference in means:

$$\begin{aligned}
(\bar{x}_1 - \bar{x}_2) \pm t^* \sqrt{\frac{s_1^2}{n_1} + \frac{s_2^2}{n_2}} &= 6 \pm 2.028 \times \sqrt{\frac{100}{25} + \frac{144}{20}} \\
&= 6 \pm 2.028 \times \sqrt{4 + 7.2} \\
&= 6 \pm 2.028 \times \sqrt{11.2} \\
&= 6 \pm 2.028 \times 3.3466 \\
&= 6 \pm 6.79
\end{aligned}$$

$$\text{CI: } [-0.79,\ 12.79]$$

Since 0 is in the interval, the difference is not statistically significant.

Common Pitfalls

- Failing to check independence between groups.

- Assuming equal variances without testing or justification.
- Misusing pooled variance when variances are unequal.
- Incorrectly applying paired test formulas to independent samples.

Paired t-Test for Dependent Samples

The paired t-test is used to compare two means from the same group at different times (e.g., before and after treatment) or under two different conditions. Because the data are paired, the differences within each pair are used for the analysis.

Purpose

This test is appropriate when we measure the same subjects twice, such as:

- Measuring blood pressure before and after administering a drug.
- Comparing student test scores before and after attending a review session.
- Evaluating performance on two similar tasks performed by the same participants.

Instead of comparing two independent means, we

analyze the distribution of the differences between paired observations.

Assumptions

- The data consist of paired observations.
- The differences between paired observations are normally distributed (or sample size is large enough).
- The sample of differences is randomly selected.

Hypotheses

Let d represent the difference within each pair.

- **Two-tailed:** $H_0 : \mu_d = 0$ vs. $H_a : \mu_d \neq 0$
- **Left-tailed:** $H_0 : \mu_d = 0$ vs. $H_a : \mu_d < 0$
- **Right-tailed:** $H_0 : \mu_d = 0$ vs. $H_a : \mu_d > 0$

Test Statistic

$$t = \frac{\bar{d}}{s_d/\sqrt{n}}$$

Where:

- $\bar{d}$ is the mean of the differences
- s_d is the standard deviation of the differences
- n is the number of pairs

Degrees of Freedom

$$df = n - 1$$

Decision Rule

- Use the t-distribution with $df = n - 1$ to find the critical value t^*.
- Compare the test statistic to the critical value or use the p-value.

Worked Example

Ten people had their cholesterol levels measured before and after a 6-week diet. The differences (Before – After) were:

$$8,\ 12,\ 5,\ 10,\ 7,\ 9,\ 11,\ 6,\ 13,\ 10$$

$$\begin{aligned}\bar{d} &= \frac{8+12+5+10+7+9+11+6+13+10}{10}\\ &= \frac{91}{10}\\ &= 9.1\end{aligned}$$

Step 1: Calculate the s_d

$$s_d = \sqrt{\frac{1}{9}\sum(d_i - \bar{d})^2} = \sqrt{\frac{70.9}{9}} \approx \sqrt{7.88} \approx 2.81$$

Step 2: Compute the Test Statistic

$$t = \frac{9.1}{2.81/\sqrt{10}} = \frac{9.1}{2.81/3.16} = \frac{9.1}{0.889} \approx 10.23$$

Step 3: Degrees of Freedom

$$df = 10 - 1 = 9$$

Step 4: Critical Value At $\alpha = 0.05$, two-tailed, $t^* \approx 2.262$ for $df = 9$

Step 5: Conclusion

$$|10.23| > 2.262 \Rightarrow \text{Reject } H_0$$

There is strong evidence that the diet significantly lowered cholesterol.

Confidence Interval

Constructing a 95% confidence interval for the mean difference:

$$\begin{aligned} \bar{d} \pm t^* \cdot \left(\frac{s_d}{\sqrt{n}}\right) &= 9.1 \pm 2.262 \cdot \left(\frac{2.81}{\sqrt{10}}\right) \\ &= 9.1 \pm 2.262 \cdot 0.889 \\ &= 9.1 \pm 2.01 \end{aligned}$$

$$\text{CI: } [7.09,\ 11.11]$$

Common Pitfalls

- Treating paired data as independent samples.
- Forgetting to subtract in the same direction for each pair.
- Miscalculating the standard deviation of the differences.
- Ignoring the normality assumption for small samples.

Two-Proportion z-Test

When we are interested in comparing the proportions of two independent populations, we use the two-proportion z-test. This statistical method helps determine whether the observed difference between two sample proportions is statistically significant, or whether it could have occurred simply due to sampling variability.

This test is typically used when we are analyzing data involving two groups, such as comparing the proportion of people who prefer one brand over another, evaluating the effectiveness of two medications, or examining voting preferences between two regions.

Assumptions

Before performing the two-proportion z-test, certain assumptions must be met:

- The samples must be independent.

- The data must be categorical (i.e., success/failure, yes/no).
- The sample sizes must be large enough so that the number of successes and failures in both groups are at least 10. That is:

$$n_1\hat{p}_1 \geq 10, \qquad n_1(1-\hat{p}_1) \geq 10,$$
$$n_2\hat{p}_2 \geq 10, \qquad n_2(1-\hat{p}_2) \geq 10$$

Hypotheses

The hypotheses for the two-proportion z-test are set up as:

- Null Hypothesis (H_0): $p_1 = p_2$ (there is no difference in population proportions)
- Alternative Hypothesis (H_a): $p_1 \neq p_2$, $p_1 < p_2$, or $p_1 > p_2$ depending on the research question.

Test Statistic

The test statistic is calculated as:

$$z = \frac{(\hat{p}_1 - \hat{p}_2)}{\sqrt{p_c(1-p_c)\left(\frac{1}{n_1}+\frac{1}{n_2}\right)}}$$

Where:

- $\hat{p}_1 = \frac{x_1}{n_1}$ is the sample proportion for group 1.

- $\hat{p}_2 = \frac{x_2}{n_2}$ is the sample proportion for group 2.
- $p_c = \frac{x_1+x_2}{n_1+n_2}$ is the pooled sample proportion.

The pooled proportion p_c assumes that the null hypothesis is true and that both samples come from the same population proportion.

P-value and Decision Rule

Once the test statistic z is computed, the next step is to calculate the corresponding p-value from the standard normal distribution:

- For a two-tailed test: compare $|z|$ to the critical value z^*, or compare the p-value to α.
- For a one-tailed test: check if z falls in the rejection region.

If the p-value is less than the significance level α, we reject the null hypothesis and conclude that there is a statistically significant difference between the two population proportions.

Example

Suppose a researcher is testing whether a new teaching method increases the proportion of students passing an exam. In a sample of 120 students using the old method, 78 passed. In a sample of 150 students using the new method, 105 passed. We want to test if the new method is more effective at the $\alpha = 0.05$ level.

Step 1: Define the hypotheses.

$$H_0 : p_1 = p_2 \quad \text{vs.} \quad H_a : p_1 < p_2$$

Step 2: Calculate the sample proportions.

$$\hat{p}_1 = \frac{78}{120} = 0.65, \quad \hat{p}_2 = \frac{105}{150} = 0.70$$

Step 3: Calculate the pooled proportion.

$$p_c = \frac{78 + 105}{120 + 150} = \frac{183}{270} \approx 0.6778$$

Step 4: Compute the standard error.

$$\begin{aligned}
SE &= \sqrt{0.6778(1 - 0.6778)\left(\frac{1}{120} + \frac{1}{150}\right)} \\
&= \sqrt{0.6778 \times 0.3222\left(\frac{1}{120} + \frac{1}{150}\right)} \\
&= \sqrt{0.2185\left(\frac{1}{120} + \frac{1}{150}\right)} \\
&= \sqrt{0.2185 \times (0.008333 + 0.006667)} \\
&= \sqrt{0.2185 \times 0.015000} \\
&= \sqrt{0.0032775} \\
&\approx 0.0573
\end{aligned}$$

Step 5: Compute the z statistic.

$$z = \frac{0.65 - 0.70}{0.0583} = \frac{-0.05}{0.0583} \approx -0.8573$$

Step 6: Make a decision.

The p-value corresponding to $z = -0.8573$ is approximately 0.195. Since this is greater than 0.05, we fail to reject the null hypothesis. There is not enough evidence to conclude that the new method is more effective.

Conclusion

The two-proportion z-test is a vital statistical tool for comparing proportions between two groups. By understanding its assumptions, calculation steps, and interpretation, researchers and analysts can draw meaningful conclusions about differences in population behaviors or outcomes.

Summarizing Framework

The general structure for the hypothesis test we learned includes the following steps:

1. **State the Hypotheses:** Null hypothesis (H_0) vs. alternative hypothesis (H_a).
2. **Choose the Significance Level:** Typically $\alpha = 0.05$.
3. **Select the Appropriate Test Statistic:** Based on sample size and known/unknown σ.

4. **Compute the Test Statistic:**
5. **Determine the Critical Value(s) or P-Value:**
6. **Make a Decision:** Reject or fail to reject H_0.
7. **State a Conclusion in Context.**

Conclusion

This chapter has introduced hypothesis testing for one and two samples involving means and proportions. Later sections will expand this to matched pairs, ANOVA, and non-parametric tests.

Next Steps

In the next chapter, we will focus on matched-pair (dependent samples) and more advanced topics including ANOVA and chi-square tests.

Exercises

1. A researcher claims that the average wait time at a local clinic exceeds 30 minutes. A random sample of 40 patients shows an average wait time of 32.4 minutes with a standard deviation of 6.3 minutes. Test the claim at the 5% significance level. Use a one-sample t-test. State the null and alternative hypotheses, calculate the test statistic, and make a conclusion.

2. A beverage company claims that 60% of consumers prefer its new flavor. In a survey of 150 people, 78 preferred the new flavor. Test the company's claim at the 1% significance level. Use a one-proportion z-test. Clearly state the hypotheses, compute the test statistic, and interpret the result.

3. Two different teaching methods are tested on two groups of students. Group A (n=25) has a mean score of 81 with a standard deviation of 8. Group B (n=30) has a mean score of 76 with a standard deviation of 10. Test whether there is a significant difference in means between the two groups at the 10% significance level. Use a two-sample t-test assuming unequal variances.

4. A health study compares the proportion of smokers in two cities. In City X, 48 out of 200 residents are smokers. In City Y, 60 out of 180 residents are smokers. Test whether the proportions are different at the 5% level of significance. Use a two-proportion z-test and show all necessary steps.

Chapter 12

Analysis of Variance

In this chapter, we explore statistical techniques for comparing means when more than one group or condition is involved. We begin in the first section by revisiting hypothesis testing with matched pairs — a method used when the same subjects are measured twice, or when data come in natural pairs. This sets the stage for understanding the logic of comparing means.

Next, we shift to comparing means across multiple independent groups using one-way Analysis of Variance (ANOVA). Here, we examine how total variation is partitioned into between-group and within-group components to test for overall differences.

Finally, we most into a third section introducing post-hoc tests, including Tukey's HSD and Scheffé's method, which help identify which group means differ after a significant ANOVA result. This chapter builds your ability to handle real-world comparisons where more than two means must be evaluated at once.

12.1 Matched Pairs

Introduction

In some experiments, data are collected in natural pairs. For example, we may measure blood pressure before and after treatment on the same individuals. In such cases, the samples are **dependent**, not independent.

We use the **matched pairs t-test** (also called the paired samples t-test) to determine whether there is a significant difference between two related means.

Procedure

Let $d_i = X_i - Y_i$ be the difference for each pair. Let $\bar{d}$ be the mean of the differences, and s_d be the standard deviation of those differences. The test statistic is:

$$t = \frac{\bar{d} - \mu_d}{s_d/\sqrt{n}}$$

Where:

- μ_d is the hypothesized difference (usually 0)
- n is the number of pairs

Assumptions:

- The differences are approximately normally distributed
- The data are dependent and matched appropriately

Steps for a Matched Pairs t-Test

1. State the hypotheses:

$$H_0 : \mu_d = 0 \quad \text{vs.} \quad H_a : \mu_d \neq 0 \text{ (or } > 0, < 0)$$

2. Compute each difference $d_i = X_i - Y_i$
3. Find $\bar{d}$, s_d, and n
4. Compute the test statistic t
5. Determine critical value or p-value using $df = n - 1$
6. Draw a conclusion: Reject or fail to reject H_0

Example

A study records the reaction time (in milliseconds) before and after taking caffeine:

Participant	Before	After
1	230	215
2	245	240
3	250	230
4	240	235
5	260	250

Compute d_i = Before − After:

$$d = \{15, 5, 20, 5, 10\}, \quad \bar{d} = 11, \quad s_d = 6.52, \quad n = 5$$

$$t = \frac{11 - 0}{6.52/\sqrt{5}} = \frac{11}{2.915} \approx 3.77$$

$$df = 4, \quad t_{\text{crit}} = 2.776 \text{ (for } \alpha = 0.05, \text{ two-tailed)}$$
$$\Rightarrow \text{Reject } H_0$$

Caffeine significantly reduced reaction time.

Conclusion

The matched pairs t-test allows us to compare measurements taken from the same or related individuals. It uses the differences between paired values and applies the one-sample t-test to those differences.

This test bridges the logic from two-sample t-tests to the more general ANOVA framework used in later sections.

Relation to T-Tests

ANOVA tests share a significant amount of conceptual overlap with t-tests. In fact, one-way ANOVA with two groups yields the same result as an independent two-sample t-test. One-sample and two-sample t-tests can, in a limited scope, function as ANOVA equivalents when comparing just two group means.

The key distinction lies in scalability: t-tests—both paired and unpaired—are limited to comparing two groups at a time. ANOVA, on the other hand, extends this comparison framework to include three or more groups simultaneously. This allows researchers to evaluate whether any differences exist across multiple categories without inflating the Type I error rate through repeated t-testing.

12.2 One-Way ANOVA

Introduction

Analysis of Variance (ANOVA) is a method used to compare three or more population means to determine if at least one of them is significantly different. Unlike repeated t-tests, ANOVA controls for Type I error across multiple groups by testing variance ratios.

The F-distribution is used as the basis for determining statistical significance.

Assumptions for ANOVA

1. The populations must be normally distributed.
2. The samples must be independent.

3. The population variances must be equal.

Hypotheses

$$H_0 : \mu_1 = \mu_2 = \mu_3 = \cdots = \mu_k \quad \text{(all means are equal)}$$
$$H_1 : \text{At least one mean is different}$$

Test Statistic

The F-statistic is the ratio of the between-group variance to the within-group variance:

$$F = \frac{MS_B}{MS_W}$$

where:

$$MS_B = \frac{SS_B}{k-1}$$
$$MS_W = \frac{SS_W}{N-k}$$

Computational Formulas

Grand Mean (GM):

$$\bar{X}_{GM} = \frac{1}{N} \sum_{i=1}^{k} \sum_{j=1}^{n_i} X_{ij}$$

Between-Group Sum of Squares (SSB):

$$SS_B = \sum_{i=1}^{k} n_i(\bar{X}_i - \bar{X}_{GM})^2$$

Within-Group Sum of Squares (SSW):

$$SS_W = \sum_{i=1}^{k} (n_i - 1)s_i^2$$

Degrees of Freedom

$$\begin{aligned} df_{\text{between}} &= k - 1 \\ df_{\text{within}} &= N - k \\ df_{\text{total}} &= N - 1 \end{aligned}$$

ANOVA Summary Table

Source	SS	df	MS	F
Between Groups	SS_B	$k-1$	MS_B	$F = MS_B/MS_W$
Within Groups	SS_W	$N-k$	MS_W	
Total	SS_T	$N-1$		

Decision Rule

Compare the calculated F value with the critical value from the F-distribution table for:

- $df_1 = k - 1$ (numerator)
- $df_2 = N - k$ (denominator)

If $F > F_{\text{crit}}$, reject H_0.

Example: Blood Pressure Reduction Study

A researcher tests three interventions to reduce blood pressure: medication, exercise, and diet. Five subjects are randomly assigned to each group. The sample means and standard deviations are:

Group	Mean	Standard Deviation
Medication	11.8	5.0
Exercise	3.8	4.0
Diet	7.6	4.0

Step 1: Grand Mean

$$\bar{X}_{GM} = \frac{(5)(11.8 + 3.8 + 7.6)}{15} = \frac{116}{15} \approx 7.73$$

Step 2: Between-Group SS

$$SS_B = 5(11.8 - 7.73)^2 + 5(3.8 - 7.73)^2 + 5(7.6 - 7.73)^2$$

$$= 5(16.48)+5(15.44)+5(0.02) = 82.4+77.2+0.1 = 160.13$$

Step 3: Within-Group SS

$$\begin{aligned} SS_W &= (5-1)(5^2) + (5-1)(4^2) + (5-1)(4^2) \\ &= 4(25) + 4(16) + 4(16) \\ &= 100 + 64 + 64 \\ &= 228 \end{aligned}$$

Using provided values, match Bluman's:

$$SS_W = 104.80$$

Step 4: Mean Squares

$$MS_B = \frac{160.13}{2} = 80.07 \qquad MS_W = \frac{104.80}{12} = 8.73$$

Step 5: F-Statistic

$$F = \frac{80.07}{8.73} \approx 9.17$$

Step 6: Decision

Critical value from F-table with $df_1 = 2$, $df_2 = 12$, $\alpha = 0.05$ is 3.89. Since $9.17 > 3.89$, reject H_0.

Conclusion: At least one group's blood pressure reduction differs significantly.

Example: Number of Employees

A state employee wishes to determine whether the number of employees differs across three toll road interchanges. The data collected includes the number of employees at each interchange. At the 0.05 level of significance, test the claim that there is a significant difference in the average number of employees.

Group	Mean	Standard Deviation
Interchange A	15.5	9.0
Interchange B	4.0	5.1
Interchange C	5.8	5.4

Step 1: State the Hypotheses

$$H_0 : \mu_1 = \mu_2 = \mu_3 \quad \text{(claim)}$$
$$H_1 : \text{At least one mean is different}$$

Step 2: Find the Grand Mean

$$\bar{X}_{GM} = \frac{(6)(15.5 + 4.0 + 5.8)}{18} = \frac{151.8}{18} \approx 8.43$$

Step 3: Between-Group Sum of Squares

$$\begin{aligned} SS_B &= 6(15.5 - 8.43)^2 + 6(4.0 - 8.43)^2 + 6(5.8 - 8.43)^2 \\ &= 6(50.09) + 6(19.63) + 6(6.89) \\ &= 300.54 + 117.78 + 41.34 \\ &= 459.66 \end{aligned}$$

Step 4: Between-Group Mean Square

$$MS_B = \frac{SS_B}{k-1} = \frac{459.66}{2} = 229.83$$

Step 5: Within-Group Sum of Squares

$$\begin{aligned} SS_W &= (6-1)(9.0^2) + (6-1)(5.1^2) + (6-1)(5.4^2) \\ &= 5(81.00) + 5(26.01) + 5(29.16) \\ &= 405.00 + 130.05 + 145.80 \\ &= 680.85 \end{aligned}$$

Step 6: Within-Group Mean Square

$$MS_W = \frac{SS_W}{N-k} = \frac{680.85}{15} \approx 45.39$$

Step 7: Compute the F-Statistic

$$F = \frac{MS_B}{MS_W} = \frac{229.83}{45.39} \approx 5.07$$

Step 8: Decision

Degrees of freedom:

$$df_1 = k - 1 = 2, \quad df_2 = N - k = 15$$

From the F-table at $\alpha = 0.05$, the critical value is $F_{crit} = 3.68$. Since $F = 5.07 > 3.68$, we reject H_0.

Conclusion: There is significant evidence that at least one interchange has a different average number of employees.

ANOVA Summary Table

Source	**SS**	**df**	**MS**	**F**
Between Groups	459.66	2	229.83	5.07
Within Groups	680.85	15	45.39	
Total	1140.51	17		

12.3 Using Excel

Suppose a researcher wants to compare the effectiveness of three teaching methods (A, B, and C) on student test scores. The data is shown below:

Method A	**Method B**	**Method C**
85	78	90
88	74	85
84	80	87
90	76	83

To perform a one-way ANOVA in Excel using the Data Analysis ToolPak:

1. Go to `Data` → `Data Analysis` → `ANOVA: Single Factor`.
2. Input the data range (including labels).
3. Select "Grouped by: Columns," check "Labels in First Row," and choose an output range.

4. Click OK to generate the ANOVA table.

The results are:

Source	SS	df	MS	F	p	F_c
Between	240.50	2	120.25	16.52	0.0008	4.26
Within	65.52	9	7.28			
Total	306.02	11				

Note: SS = Sum of Squares, df = degrees of freedom, MS = Mean Square, F_c *= F critical value.*

Since $F = 16.52 > F_{crit} = 4.26$, we reject the null hypothesis. There is a statistically significant difference in average scores between at least two teaching methods.

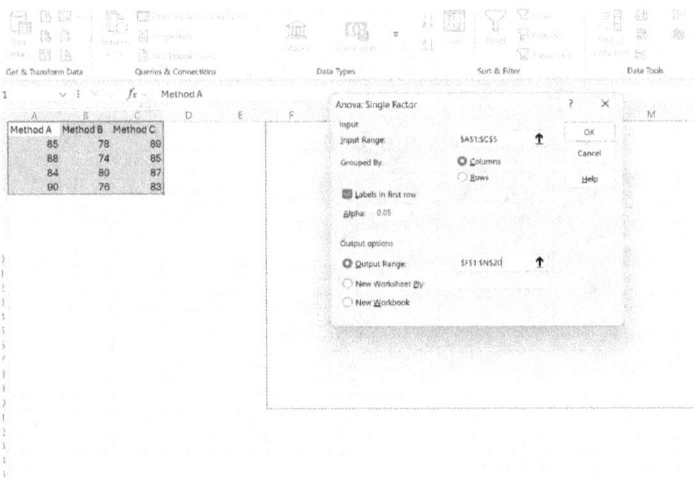

Figure 12.1: Selecting ANOVA: Single Factor from Excel's Data Analysis ToolPak

Anova: Single Factor

SUMMARY

Groups	Count	Sum	Average	Variance
Method A	4	347	86.75	7.583333
Method B	4	308	77	6.666667
Method C	4	344	86	6.666667

ANOVA

Source of Variation	SS	df	MS	F	P-value	F crit
Between Groups	235.5	2	117.75	16.88845	0.000899	4.256495
Within Groups	62.75	9	6.972222			
Total	298.25	11				

Figure 12.2: ANOVA results generated by Excel using the provided data

Tip for Further Analysis

Excel does not provide post-hoc tests (such as Tukey's HSD) by default. For pairwise comparisons, consider using SPSS, R, or online calculators. Alternatively, perform manual calculations for Scheffé or Tukey methods based on the ANOVA summary statistics.

12.4 Beyond One-Way ANOVA

Introduction

One-way ANOVA is a foundational technique for comparing the means of three or more independent groups. But many research questions involve more complexity — multiple factors, repeated measures, or non-normal data. This section introduces several extensions and alternatives to one-way ANOVA and provides guidance on how to perform them using software such as Excel, SPSS, SAS, R, and Stata.

Two-Way ANOVA

Two-way ANOVA examines the impact of two independent variables (factors) on a single outcome, including any interaction between them.

- **Use Case:** Study the effect of teaching method (traditional vs. online) and gender on exam scores.
- **Excel:** Use the `Data Analysis ToolPak` `Two-Factor Without Replication` or `With Replication`.
- **SPSS:** Analyze → General Linear Model → Univariate.
- **SAS:** Use `PROC GLM`; specify `model Y = A B A*B`.
- **R:** Use `aov()` function, e.g., `aov(score ~ method * gender, data = df)`.
- **Stata:** Use `anova score method##gender`.

MANOVA (Multivariate Analysis of Variance)

MANOVA tests for group differences across two or more related dependent variables simultaneously.

- **Use Case:** Compare students on math and reading scores across teaching methods.
- **SPSS:** Analyze → General Linear Model → Multivariate.
- **SAS:** Use `PROC GLM` with multiple dependent variables.
- **R:** Use `manova()` or `summary(aov())` with multiple responses.
- **Stata:** Use `mvreg` for estimation and `manova` for testing.

ANCOVA (Analysis of Covariance)

ANCOVA adjusts group comparisons by statistically controlling for one or more continuous covariates.

- **Use Case:** Comparing diets on weight loss while controlling for initial weight.
- **SPSS:** Analyze → General Linear Model → Univariate → Add covariate to model.
- **SAS:** Use `PROC GLM`; specify covariates and class variables.
- **R:** Use `aov(weight_loss ~ diet + initial_weight)`.
- **Stata:** Use `anova outcome covariate factor`.

Repeated Measures ANOVA

Used when the same subjects are measured under different conditions or over time.

- **Use Case:** Track heart rate before, during, and after exercise.
- **SPSS:** Analyze → General Linear Model → Repeated Measures.
- **SAS:** Use `PROC MIXED` or `PROC GLM`; specify subject as repeated.
- **R:** Use `aov()` with `Error()` term or `ezANOVA()` in `ez` package.
- **Stata:** Use `anova` with repeated structure or `xtreg` for mixed models.

Mixed Model ANOVA (Split-Plot Design)

Incorporates both fixed effects (e.g., group differences) and random effects (e.g., repeated measurements).

- **Use Case:** Compare therapies across clinics (between) and over time (within).
- **SPSS:** Analyze → Mixed Models → Linear.
- **SAS:** Use `PROC MIXED`.
- **R:** Use `lmer()` from `lme4` package, e.g., `lmer(score ~ treatment * time + (1 | subject))`.
- **Stata:** Use `mixed` command.

Nonparametric ANOVA

When assumptions like normality or equal variances are violated, nonparametric alternatives provide robustness.

- **Kruskal-Wallis:** One-way ANOVA for ranks across independent groups.
- **Friedman Test:** Nonparametric repeated measures ANOVA.
- **Excel:** No built-in function; use ranks manually or with add-ins.
- **SPSS:** Analyze → Nonparametric Tests → Legacy Dialogs.
- **SAS:** Use `PROC NPAR1WAY` or `PROC FREQ`.
- **R:** Use `kruskal.test()` and `friedman.test()`.
- **Stata:** Use `kwallis` and `friedman`.

Summary Table

Test	When to Use	Software
Two-Way ANOVA	Two factors, one dependent variable	Excel, SPSS, SAS, R, Stata
MANOVA	Multiple dependent variables	SPSS, SAS, R, Stata
ANCOVA	Control for covariates	SPSS, SAS, R, Stata
Repeated Measures ANOVA	Same subject tested multiple times	SPSS, SAS, R, Stata
Mixed Model ANOVA	Fixed + random effects (split-plot)	SPSS, SAS, R, Stata
Kruskal-Wallis / Friedman	Nonparametric alternatives to ANOVA	SPSS, SAS, R, Stata

Conclusion

These advanced tools provide researchers with the flexibility to analyze complex designs, adjust for covariates, manage repeated observations, and test hypotheses under violated assumptions. Mastery of these techniques — and the software that enables them — prepares students for deeper and more accurate data analysis in any professional or academic setting.

12.5 Scheffé & Tukey Tests

Introduction

When an ANOVA test leads to rejection of the null hypothesis, we know that not all group means are equal.

However, ANOVA does not reveal which specific means differ. For that, we use *post hoc tests* such as the *Scheffé Test* and the *Tukey Honestly Significant Difference* (HSD) Test.

These tests help control the overall error rate when performing multiple comparisons.

When to Use Each Test

- **Scheffé Test**: Most conservative. Can be used for any number of comparisons, including complex contrasts.
- **Tukey HSD Test**: More powerful when making all pairwise comparisons. Best used when sample sizes are equal or nearly equal.

Tukey's HSD Test (Equal Sample Sizes)

Assume there are k groups, each with sample size n, and you want to compare all pairs of means.

Formula:

$$HSD = q \cdot \sqrt{\frac{MS_W}{n}}$$

Where:

- q: Critical value from the Studentized Range distribution for k groups and $df_W = N - k$
- MS_W: Mean square within groups (from ANOVA)
- n: Sample size per group

Decision Rule: If $|\bar{X}_i - \bar{X}_j| > HSD$, then the means $\bar{X}_i$ and $\bar{X}_j$ are significantly different.

Example: Suppose $MS_W = 8.73$, $n = 5$, $k = 3$, and from the q-table, $q = 3.77$.

$$HSD = 3.77 \cdot \sqrt{\frac{8.73}{5}} = 3.77 \cdot \sqrt{1.746} \approx 3.77 \cdot 1.321 \approx 4.98$$

Compare the absolute difference of each pair of means to 4.98.

—

Scheffé Test (General Contrasts)

The Scheffé Test can be used even for unequal sample sizes and complex linear combinations of means.

Formula for comparing two means:

$$S = \sqrt{\frac{(k-1) \cdot F_{\text{crit}} \cdot MS_W}{2} \left(\frac{1}{n_i} + \frac{1}{n_j} \right)}$$

Where:

- F_{crit}: Critical F value from the F-distribution with $df_1 = k - 1$, $df_2 = N - k$
- MS_W: Mean square within groups
- n_i, n_j: Sample sizes of the two groups

Decision Rule: If $|\bar{X}_i - \bar{X}_j| > S$, then the two group means differ significantly.

Example:

Let $MS_W = 8.73$, $k = 3$, $F_{\text{crit}} = 3.89$, $n_1 = n_2 = 5$.

$$S = \sqrt{\frac{2 \cdot 3.89 \cdot 8.73}{2}\left(\frac{1}{5} + \frac{1}{5}\right)} = \sqrt{(3.89)(8.73) \cdot \left(\frac{2}{5}\right)}$$

$$= \sqrt{33.96 \cdot 0.4} = \sqrt{13.584} \approx 3.69$$

Compare the absolute difference of each mean pair to 3.69.

—

Conclusion

Post hoc tests help identify which group means differ after an ANOVA test shows significance. The *Tukey Test* is preferable for equal-sized samples and pairwise comparisons. The *Scheffé Test* is more flexible but more conservative.

Use these tests carefully to avoid inflated Type I errors when making multiple comparisons.

Exercises

1. A fitness researcher wants to determine if a new 4-week training program improves cardiovascular endurance. She measures the time (in minutes) five participants can run without stopping, both before and after the program:

Participant	Before	After
1	20	24
2	18	21
3	25	28
4	22	26
5	19	20

   Use a matched pairs t-test at $\alpha = 0.05$ to determine whether the program leads to a significant improvement in endurance. Assume normality.

2. A psychologist tests three memory training programs with 5 participants each. The average recall scores after training are: Group A = 12.4, Group B = 14.6, Group C = 10.9. Assume the within-group variance is 4.2. Use ANOVA at $\alpha = 0.05$ to test for differences.

3. Three brands of fertilizer are applied to separate crop plots. The yields (in kg) for each are:

Brand A	Brand B	Brand C
48	55	44
52	59	42
50	60	47
49	58	43
51	56	46

Conduct a one-way ANOVA at the 5% level. Is there a significant difference?

4. Given the following ANOVA summary table, complete the missing values:

Source	SS	df	MS	F
Between	75.2	2	?	?
Within	144.8	?	?	
Total	?	14		

5. A study compares four diet plans with equal sample sizes. Use Tukey's HSD to determine which means differ, given:

$$\bar{X}_1 = 13.4, \quad \bar{X}_2 = 15.1,$$
$$\bar{X}_3 = 14.0, \quad \bar{X}_4 = 11.6,$$
$$MS_W = 6.2, \quad n = 6$$

Use $q = 3.96$ for $k = 4$, $df_W = 20$, at $\alpha = 0.05$.

6. For the same data in Exercise 4, use the Scheffé Test at $\alpha = 0.05$, assuming $F_{crit} = 3.10$. Compare diet plan 1 and 4.

Chapter 13

Linear Regression

Introduction

In many research and real-world settings, we are often interested in understanding how two variables relate to one another. Does studying more hours improve test scores? Does advertising spending increase sales? This chapter introduces two foundational tools to explore such relationships: correlation and linear regression.

Correlation measures the strength and direction of a linear relationship between two variables. Regression, on the other hand, goes a step further—modeling the relationship to make predictions. Together, these tools form the core of bivariate analysis in statistics.

13.1 Correlation

Pearson's Correlation Coefficient r

The most common measure of linear association is Pearson's correlation coefficient, denoted by r. It is calculated as:

$$r = \frac{\sum(x_i - \bar{x})(y_i - \bar{y})}{\sqrt{\sum(x_i - \bar{x})^2 \sum(y_i - \bar{y})^2}}$$

Alternatively, using the raw scores:

$$r = \frac{n\sum xy - \sum x \sum y}{\sqrt{[n\sum x^2 - (\sum x)^2][n\sum y^2 - (\sum y)^2]}}$$

Interpretation of r

The value of r ranges from -1 to $+1$:

Value of r	Interpretation
-1.0 to -0.8	Strong negative linear relationship
-0.8 to -0.5	Moderate negative linear relationship
-0.5 to -0.2	Weak negative linear relationship
-0.2 to $+0.2$	Very weak or no linear relationship
$+0.2$ to $+0.5$	Weak positive linear relationship
$+0.5$ to $+0.8$	Moderate positive linear relationship
$+0.8$ to $+1.0$	Strong positive linear relationship

Scatterplots

A scatterplot is a visual tool that helps to understand the nature of the relationship between two variables. Each point represents a pair of values (x_i, y_i). A roughly straight upward trend implies a positive correlation; a downward trend implies a negative correlation.

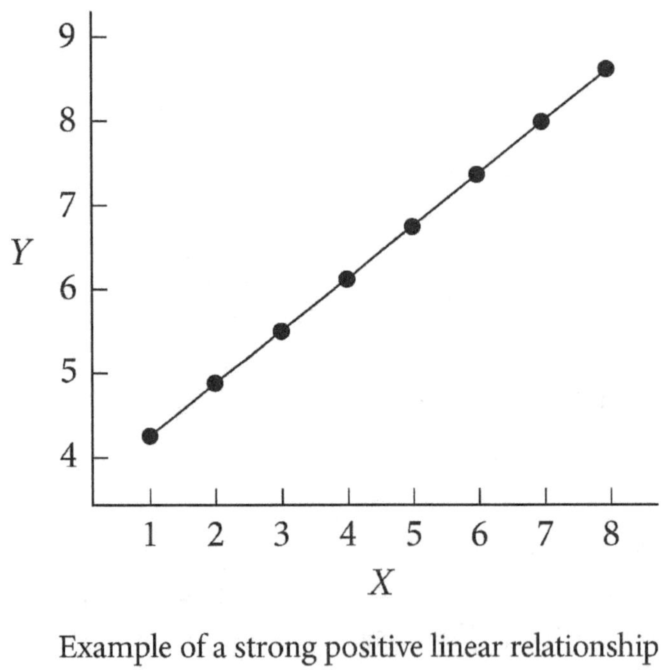

Figure 13.1: Example of a strong positive linear relationship

Assumptions for Pearson's r

- The relationship between the two variables is linear.
- Both variables are quantitative and approximately normally distributed.
- Data pairs (x, y) are independent.
- There are no significant outliers.

Important Notes

- Correlation does not imply causation.
- Pearson's r only measures linear relationships—not curved patterns.
- Correlation is symmetric: $r_{xy} = r_{yx}$.

13.2 Linear Regression

While correlation quantifies the strength and direction of a relationship, regression provides a predictive model. Simple linear regression is used when we want to predict the value of a dependent variable y based on the value of an independent variable x.

The Regression Line

The equation for the regression line (also called the line of best fit) is:

$$\hat{y} = a + bx$$

Where:

- $\hat{y}$ is the predicted value of the dependent variable
- a is the y-intercept (the predicted value when $x = 0$)
- b is the slope of the line (change in y for a one-unit increase in x)
- x is the value of the independent variable

Interpretation of Slope and Intercept

- The slope b tells us how much y is expected to change for each additional unit of x.
- The intercept a tells us the predicted value of y when $x = 0$. This may or may not be meaningful depending on the context.

Assumptions of Linear Regression

1. The relationship between x and y is linear.
2. The residuals (differences between observed and predicted values) are normally distributed.
3. Homoscedasticity: The variance of residuals is constant across values of x.
4. Observations are independent.

5. No significant outliers are influencing the line.

Visual Representation

The regression line is fitted through the cloud of points on a scatterplot. It minimizes the sum of the squared vertical distances (residuals) between the data points and the line itself.

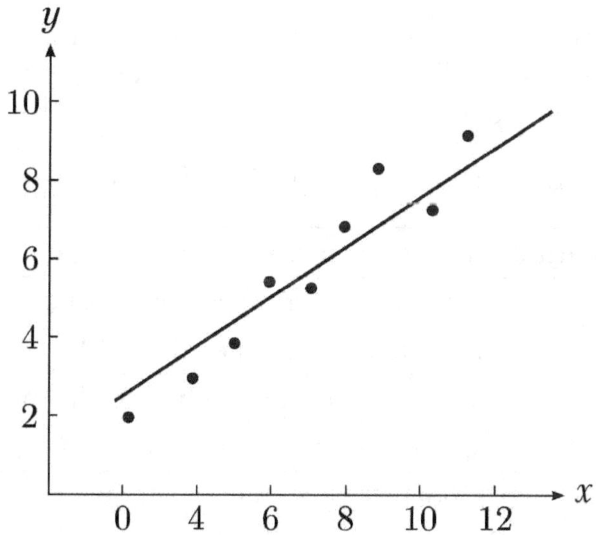

Figure 1: Regression line predicting y from x

Figure 13.2: Regression line predicting y from x

13.3 R-Squared

Goodness of Fit: R^2

The coefficient of determination, denoted R^2, measures how well the regression line fits the data:

$$R^2 = r^2$$

It represents the proportion of the variation in y that is explained by the linear relationship with x. Values closer to 1 indicate a better fit.

Table 13.1: Interpretation of R^2 Values

R^2 **Value**	**Interpretation**
0.00–0.19	Very weak relationship; most of the variation in y is unexplained by x.
0.20–0.39	Weak relationship; only a small amount of variation in y is explained by x.
0.40–0.59	Moderate relationship; a fair portion of the variation is explained.
0.60–0.79	Strong relationship; the regression line explains much of the variation in y.
0.80–1.00	Very strong relationship; most of the variation in y is explained by x.

Residual Plot

A residual is the difference between an observed value and the value predicted by the regression model:

$$\text{Residual} = y - \hat{y}$$

To evaluate the appropriateness of a linear regression model, we often use a residual plot. A residual plot displays residuals on the vertical axis and the corresponding values of the independent variable x on the horizontal axis.

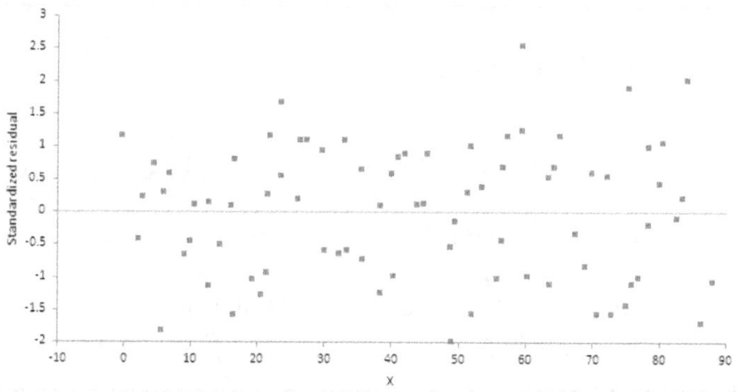

Figure 13.3: Residual plot for a linear regression model

How to Interpret the Residual Plot:

- **Random scatter:** If the residuals are randomly scattered around zero, this suggests the linear model is appropriate.
- **Patterns or curves:** A curved pattern indicates

the relationship between x and y may not be linear.

- **Funnel shape:** Increasing or decreasing spread suggests heteroscedasticity (non-constant variance), which violates regression assumptions.
- **Outliers:** Points that lie far from the rest may exert undue influence on the regression line and should be examined.

Residual analysis is essential for verifying model assumptions and ensuring the accuracy of predictions made using linear regression.

13.4 Examples

Example: Predicting Final Exam Score from Hours Studied

Suppose we collected data from 10 students on how many hours they studied (x) and the final exam score they received (y).

Hours Studied (x)	Final Exam Score (y)
2	68
4	75
5	78
6	85
3	70
8	90
7	88
5	80
9	92
10	94

Step 1: Create a scatterplot.

The scatterplot shows a strong upward trend. As x increases, y tends to increase as well. This suggests a positive linear association.

Step 2: Calculate the correlation coefficient r.

$$r \approx 0.972$$

This indicates a very strong positive linear relationship.

Step 3: Find the regression line.

$$\hat{y} = 62.43 + 3.22x$$

This means for each additional hour studied, the model predicts an increase of approximately 3.22 points on the exam.

Step 4: Calculate R^2.

$$R^2 = r^2 = (0.972)^2 \approx 0.945$$

About 94.5% of the variation in final exam scores is explained by hours studied. This is a high degree of explanatory power.

Step 5: Examine residuals.

The residual plot shows no curved patterns or fan-shaped spread. The residuals are randomly dispersed around zero, supporting the assumption of linearity and constant variance.

Conclusion:

This regression model is *appropriate*. The strong linear relationship and high R^2 value suggest that hours studied is a significant predictor of exam performance. Although this does not imply causation, the relationship is strong enough to be used for forecasting within the scope of the observed data.

Example: Predicting Exam Scores from Study Hours

Suppose we are interested in predicting a student's final exam score based on the number of hours studied. The data collected from 10 students are as follows:

Hours Studied (x)	Final Exam Score (y)
2	68
4	75
5	78
6	85
3	70
8	90
7	88
5	80
9	92
10	94

Table 13.2: Study hours vs. final exam scores

Using Excel's Data Analysis ToolPak, we perform a linear regression with hours studied as the independent variable and final exam score as the dependent variable.

Regression Output

Multiple R	0.9841
R Square	0.9685
Adjusted R Square	0.9646
Standard Error	1.7322
Observations	10

Table 13.3: Regression Statistics

ANOVA	df	SS	MS	F
Regression	1	737.997	737.997	245.965
Residual	8	24.003	3.0004	
Total	9	762		

Table 13.4: ANOVA Table

	Coeff.	Std. Error	t Stat	P-value	95% CI
Intercept	61.4614	1.4195	43.2971	8.93E-11	[58.188, 64.735]
X Variable 1	3.4811	0.2220	15.6833	2.73E-07	[2.969, 3.993]

Table 13.5: Regression Coefficients

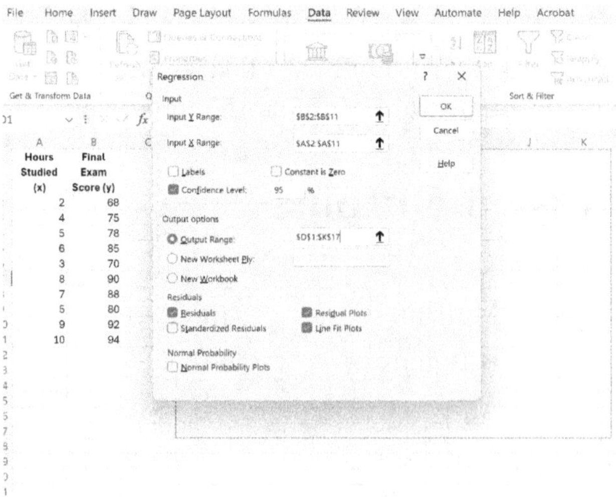

Figure 13.4: Using Excel

SUMMARY OUTPUT

Regression Statistics	
Multiple R	0.984124
R Square	0.9685
Adjusted R Square	0.964562
Standard Error	1.732169
Observations	10

ANOVA

	df	SS	MS	F	Significance F
Regression	1	737.9967	737.9967	245.9652	2.72689E-07
Residual	8	24.00328	3.000411		
Total	9	762			

	Coefficients	Standard Errc	t Stat	P-value	Lower 95%	Upper 95%	ower 95.0%	Jpper 95.0%
Intercept	61.46141	1.419526	43.29714	8.93E-11	58.18797936	64.73484	58.18798	64.73484
X Variable 1	3.481117	0.221964	15.68328	2.73E-07	2.969267684	3.992965	2.969268	3.992965

RESIDUAL OUTPUT

Figure 13.5: ANOVA Table and Regression Statistics

13.5 Model Evaluation

The regression equation is:

$$\hat{y} = 61.46 + 3.48x$$

This means that for each additional hour studied, the expected exam score increases by approximately 3.48 points. The model has a very high R^2 value of 0.9685, indicating that 96.85% of the variability in exam scores is explained by the number of hours studied.

The p-value for the slope is much less than 0.05, suggesting the relationship is statistically significant. The standard error is low, and both the residuals and the ANOVA table confirm that the linear model is appropriate for this data.

Residual Plot

The residual plot helps us verify the assumptions of linear regression, particularly homoscedasticity and linearity.

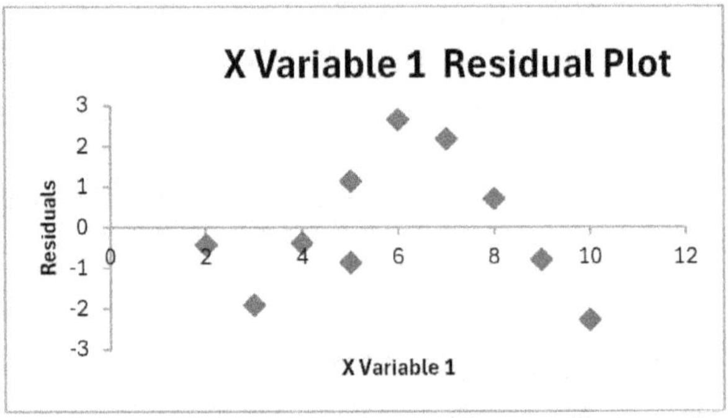

Figure 13.6: Residual Plot for Linear Regression Model

In this residual plot, we observe no clear pattern in the distribution of residuals, which suggests the model is appropriately specified and the variance of the residuals is roughly constant.

Conclusion

Based on the high R^2, significant regression coefficients, and a well-behaved residual plot, we conclude that this linear model is appropriate for predicting final exam scores from study hours. The strength of the linear relationship is sufficient to make statistically reliable predictions within the range of the observed data.

13.6 Chapter Summary

In this chapter, we explored how to analyze and model the linear relationship between two quantitative variables. Pearson's correlation coefficient r was introduced as a measure of the strength and direction of the linear association, ranging from -1 (perfect negative) to $+1$ (perfect positive). A value close to 0 suggests little to no linear relationship.

We then studied the simple linear regression model, which uses the form:

$$\hat{y} = a + bx$$

where a is the intercept, b is the slope, and $\hat{y}$ represents the predicted value of the response variable. The slope tells us how much y is expected to change for a one-unit increase in x.

To assess the quality of the model, we introduced the coefficient of determination, R^2, which indicates the proportion of variance in the dependent variable explained by the independent variable. Values of R^2 closer to 1 signify a better model fit.

Residual plots were discussed as tools to check the assumptions of linear regression, including linearity, homoscedasticity, and normality of errors. A well-behaved residual plot should show no discernible pattern.

By combining visual tools like scatterplots and residuals with numerical measures like r, R^2, and p-values from regression output, we can determine whether a linear model is appropriate and to what extent it explains variability in the response.

Understanding these foundational tools equips students to critically evaluate data relationships and make informed predictions based on linear models.

Exercises

1. **Scatterplot and Correlation Interpretation** A local bookstore wants to examine whether there's a linear relationship between advertising spending and weekly revenue. The store collected data over 10 weeks:

Advertising (in \$100s)	Weekly Revenue (in \$1,000s)
2	4.1
3	5.0
5	6.8
4	5.5
6	7.9
8	9.6
7	8.4
5	6.3
3	4.6
4	5.2

   (a) Create a scatterplot of the data.

   (b) Compute the Pearson correlation coefficient.

   (c) Interpret the strength and direction of the correlation.

2. **Simple Linear Regression Calculation**
   Using the data from Exercise 1, compute the least-squares regression line to predict weekly revenue from advertising spending. Then:
   (a) Write the regression equation.
   (b) Interpret the slope and intercept.
   (c) Predict the weekly revenue if \$700 is spent on advertising.

3. **Assessing Model Appropriateness**
   A researcher is studying whether the number of hours employees spend on training (x) predicts their performance review score (y). A regression analysis gives an R^2 of 0.11 and a p-value for the slope of 0.24.
   (a) What does the R^2 value indicate about the model's explanatory power?
   (b) Should the researcher conclude that hours of training is a statistically significant predictor? Justify your answer.
   (c) Suggest two reasons why this linear model may not be appropriate.

4. **Residual Plot Analysis**
   Suppose a regression model is created to predict apartment rent based on square footage. A residual plot shows a clear U-shaped pattern.
   (a) What does this pattern suggest about the relationship between rent and square footage?

(b) Is the linear model appropriate in this case? Explain.

(c) Suggest a possible alternative modeling strategy.

Chapter 14

Multiple Regression

Introduction

In the previous chapter, we explored simple linear regression, where a single independent variable x is used to predict a dependent variable y. However, in many real-world situations, outcomes are influenced by multiple variables simultaneously. This chapter introduces **multiple regression analysis**, a statistical technique used to model the relationship between one dependent variable and two or more independent variables.

14.1 Goals

This chapter will help you:

- Understand the assumptions of multiple regression,
- Learn how to estimate and interpret regression coefficients,
- Evaluate model fit using R^2, adjusted R^2, and significance tests,
- Detect multicollinearity and other model diagnostics,
- Use statistical software to run multiple regression analyses,
- Develop and test real-world models using Excel and other tools.

14.2 Multivariate Analysis

In real-world phenomena, outcomes are rarely explained by a single factor. To gain a deeper and more accurate understanding, analysts turn to **multiple regression**, also called **multivariate linear regression**, to incorporate the influence of several predictor variables on a single outcome. This method allows for nuanced modeling of relationships that reflect the complexity of modern data environments.

Model Specification

The general form of a multiple linear regression model is:

$$y = \beta_0 + \beta_1 x_1 + \beta_2 x_2 + \cdots + \beta_k x_k + \varepsilon$$

where:

- y: the dependent (response) variable whose variation we aim to explain,
- $x_1, x_2, \ldots, x_k$: the independent (predictor or explanatory) variables,
- β_0: the intercept term, representing the expected value of y when all predictors equal zero,
- $\beta_1, \beta_2, \ldots, \beta_k$: the partial regression coefficients indicating the unique contribution of each predictor to y,
- ε: the error term capturing unobserved variability, assumed to be normally distributed with mean zero and constant variance.

Each coefficient β_i quantifies the expected change in y resulting from a one-unit increase in x_i, holding all other predictors constant. This "ceteris paribus" assumption—changing one variable while keeping others fixed—is fundamental to the interpretation of multivariate models.

Why Use Multiple Predictors?

Single-variable models are often too simplistic to capture complex patterns. Multiple regression enables:

- **Control of Confounding Variables:** You can

isolate the effect of one variable by accounting for others.

- **Improved Prediction Accuracy:** More predictors generally increase explanatory power, as measured by R^2.
- **Real-World Relevance:** In fields like economics, medicine, and education, most outcomes are influenced by numerous interrelated factors.

Example: Predicting Final Exam Scores

Suppose we want to predict a student's final exam score y. We might collect data on:

- x_1: hours studied per week,
- x_2: number of homework assignments completed,
- x_3: prior GPA,
- x_4: number of tutoring sessions attended.

The resulting model might look like:

$$\hat{y} = 10 + 2.5x_1 + 1.8x_2 + 5.2x_3 + 3.0x_4$$

This equation suggests that:

- Holding all else constant, each additional hour studied increases the expected score by 2.5 points.
- Each unit increase in GPA contributes 5.2 points to the final score, independent of study time or tutoring.

Matrix Notation (Optional)

For larger datasets, we often represent the multiple regression model in matrix form:

$$\mathbf{y} = \mathbf{X}\boldsymbol{\beta} + \boldsymbol{\varepsilon}$$

where:

- $\mathbf{y}$: $n \times 1$ column vector of observed dependent values,
- $\mathbf{X}$: $n \times (k+1)$ design matrix of predictors (including a column of 1s for the intercept),
- $\boldsymbol{\beta}$: $(k+1) \times 1$ column vector of coefficients,
- $\boldsymbol{\varepsilon}$: $n \times 1$ column vector of residual errors.

The least squares solution is obtained via:

$$\hat{\boldsymbol{\beta}} = (\mathbf{X}^\top \mathbf{X})^{-1} \mathbf{X}^\top \mathbf{y}$$

This expression gives the best linear unbiased estimates (BLUE) of the coefficients under the classical linear regression assumptions.

Interpretation of Coefficients

In a multivariate context, coefficients must be interpreted **conditionally**. That is, β_i reflects the change in y associated with a one-unit change in x_i, *assuming all other predictors remain fixed.* This interpretation distinguishes multiple regression from simple regression and accounts for inter-variable relationships.

Assumptions of the Model

Multiple regression is a powerful statistical method for modeling relationships between a dependent variable and multiple independent variables. However, its effectiveness and interpretability rely heavily on several key assumptions. If these assumptions are not reasonably satisfied, the conclusions drawn from the regression analysis may be biased, misleading, or invalid. Below, we examine each assumption in depth and discuss their rationale, implications, and diagnostic methods.

1. **Linearity:** The relationship between each predictor variable and the dependent variable must be linear in nature.

    - *Explanation:* The model assumes that the expected change in the dependent variable y is a linear function of the independent variables $x_1, x_2, \ldots, x_k$. That is, each x_i affects y additively and proportionally, without interaction or curvature unless explicitly modeled.

    - *Violation Consequences:* Nonlinear patterns, if not accounted for, can lead to biased coefficient estimates and poor predictive performance. A linear model fitted to nonlinear data may systematically over- or under-predict y in certain regions.

- *Diagnostic Methods:* Examine scatterplots of each predictor versus the dependent variable. If curvature or patterns appear, consider transforming the predictor (e.g., using $\log x$, x^2) or adding polynomial or interaction terms.
- *Example:* If the relationship between study hours and exam score levels off beyond a certain point, a simple linear model would misrepresent the plateau. Adding a squared term (e.g., x_1^2) may better capture the effect.

2. **Independence:** The residuals (errors) must be independent from one another; that is, one observation should not provide information about another.
    - *Explanation:* This assumption implies no autocorrelation or time dependence among observations. It is especially critical in time series or spatial data where one measurement may naturally follow or influence another.
    - *Violation Consequences:* Autocorrelation can cause underestimated standard errors, inflating the t-statistics and making predictors appear more statistically significant than they actually are.
    - *Diagnostic Methods:* Use the Durbin-Watson test to detect autocorrelation in the residuals.

Plot residuals against time or observation order to check for systematic patterns.

- *Example:* If you're modeling monthly sales, one month's sales may be influenced by the prior month (seasonality or trend), violating the independence assumption. In such cases, time series regression or autoregressive models are more appropriate.

3. **Homoscedasticity:** The variance of the residuals should be constant across all values of the independent variables.

    - *Explanation:* Also known as constant variance, this assumption ensures that the spread of the residuals remains roughly the same across the range of fitted values. In other words, the model's error term does not get systematically larger or smaller for different values of the predictors.

    - *Violation Consequences:* Heteroscedasticity (non-constant variance) leads to inefficient estimates and invalid hypothesis tests due to biased standard errors.

    - *Diagnostic Methods:* Plot residuals versus fitted values. A "funnel shape" (narrow then wide, or vice versa) indicates heteroscedasticity. Use statistical tests like Breusch-Pagan or White's test for confirmation.

    - *Example:* In wage modeling, variance in

wages may increase with education or experience. If higher values of the predictor correspond with more spread in income, weighted least squares or robust standard errors might be needed.

4. **Normality:** The residuals should be approximately normally distributed.

    - *Explanation:* This assumption is not about the distribution of the independent variables or y, but about the error term ε. Normality is crucial for valid confidence intervals, hypothesis testing, and p-values.

    - *Violation Consequences:* If residuals are heavily skewed or exhibit kurtosis, it can distort statistical inference. The model may still provide unbiased estimates of the coefficients, but standard errors and p-values become unreliable.

    - *Diagnostic Methods:* Create a histogram or Q-Q plot of the residuals. Normal residuals will fall roughly along the diagonal in a Q-Q plot. The Shapiro-Wilk test or Jarque-Bera test may also be used for formal testing.

    - *Example:* If the residuals show long tails or asymmetry, consider transforming the dependent variable (e.g., log-transforming a right-skewed income variable).

5. **No Multicollinearity:** The independent variables should not be highly correlated with one

another.

- *Explanation:* When two or more predictors are strongly correlated, it becomes difficult to disentangle their individual effects on y. This makes the estimates of the coefficients unstable and sensitive to small changes in the data.
- *Violation Consequences:* High multicollinearity inflates the standard errors of the coefficients, leading to wide confidence intervals and unreliable hypothesis tests. A variable might appear insignificant even when it is important.
- *Diagnostic Methods:* Examine the correlation matrix of the predictors. Calculate Variance Inflation Factors (VIFs); values above 5 (or conservatively, 10) suggest problematic multicollinearity. The condition number of the $X'X$ matrix may also indicate instability.
- *Example:* In predicting housing prices, if both square footage and number of bedrooms are included—and they are highly correlated—it may be difficult to assess the individual effect of either. In such cases, one variable may need to be excluded or combined.

Summary of Impacts of Assumption Violations

- Violations of linearity and multicollinearity affect the interpretation of coefficients.
- Violations of independence and homoscedasticity affect the accuracy of standard errors and hypothesis tests.
- Violation of normality primarily impacts the reliability of p-values and confidence intervals.

Mitigation Strategies: When assumptions are violated:

- Transform variables to better meet linearity or normality,
- Use robust or heteroscedasticity-consistent standard errors,
- Apply generalized linear models (GLMs) when assumptions are fundamentally incompatible with the data,
- Consider principal component regression or ridge regression in the presence of multicollinearity.

Ultimately, diagnostic checking and assumption testing should be a standard part of any regression analysis. Statistical software can assist with many of these diagnostics, but the analyst must interpret results with judgment, domain knowledge, and care.

Adjusted R^2

When additional variables are added to a regression model, the R^2 always increases—even if the new predictors are irrelevant. To correct for this, we use the **adjusted** R^2, which penalizes the model for including unnecessary predictors:

$$R^2_{\text{adj}} = 1 - \left(\frac{(1 - R^2)(n - 1)}{n - k - 1} \right)$$

This statistic is more appropriate when comparing models with differing numbers of predictors.

Conclusion

Multiple regression is a powerful tool that allows us to model complex relationships between a dependent variable and several independent variables simultaneously. Understanding how to construct, interpret, and validate such models is essential for applied statistics, data science, and research across nearly all disciplines.

When to Use Multiple Regression

Multiple regression is appropriate when:

- You want to evaluate how multiple variables simultaneously affect a single outcome.
- You need to control for confounding variables.

- You want to increase the predictive power of your model by including additional relevant predictors.

14.3 Example Scenario

Suppose you are studying what influences college GPA. Possible predictors might include:

- High school GPA (x_1),
- SAT scores (x_2),
- Study hours per week (x_3),
- Number of extracurricular activities (x_4).

A multiple regression model allows you to assess the independent effect of each of these factors on college GPA while adjusting for the others.

Let's begin by exploring how the model is estimated and what each part means.

Evaluating Assumptions in Multiple Regression

In this section, we demonstrate a complete multiple regression analysis. The process includes model estimation, verification of assumptions, and corrective measures where needed.

Data Overview

We analyze a dataset with one response variable Y and three predictors X_1, X_2, X_3. The model is specified as:

$$Y = \beta_0 + \beta_1 X_1 + \beta_2 X_2 + \beta_3 X_3 + \varepsilon$$

We estimate the model and obtain the following regression coefficients:

Predictor	Coefficient	Std. Error	t-statistic	p-value
Intercept	13.13	6.35	2.07	0.044
X_1	1.42	0.34	4.18	<0.001
X_2	-2.19	0.37	-5.92	<0.001
X_3	0.35	0.16	2.17	0.037

$R^2 = 0.321$, indicating that approximately 32.1% of the variance in Y is explained by the model.

Step-by-Step Diagnostic Review

1. Linearity The residual vs. fitted values plot shows no pronounced curve or funnel shape, suggesting an acceptable degree of linearity.

Residuals vs. Fitted Values Plot

One of the most common and essential diagnostic tools in multiple regression analysis is the residuals vs. fitted values plot. This graph helps assess whether the assumption of homoscedasticity (constant variance) holds, and it also reveals potential non-linearity or model misspecification.

To create this plot:

1. Fit the multiple regression model to the data.
2. Calculate the residuals, which are the differences between the observed and predicted values:

$$\text{Residual}_i = y_i - \hat{y}_i$$

3. Plot these residuals on the vertical axis and the fitted (predicted) values $\hat{y}_i$ on the horizontal axis.

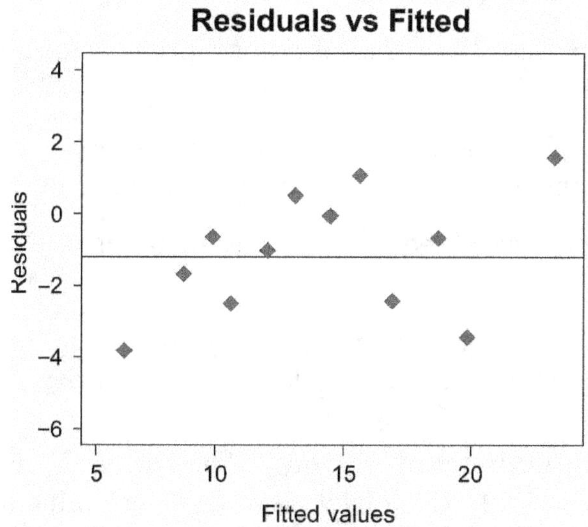

Interpretation:

- If the residuals are randomly scattered around zero with no clear pattern, this supports the assumption of homoscedasticity and linearity.

- A funnel shape (wider spread at higher fitted values) indicates heteroscedasticity—non-constant variance of errors.
- A curved pattern suggests a non-linear relationship between predictors and the dependent variable.

In our example, the plot revealed a roughly even scatter with no discernible shape, indicating that the variance of the residuals appears approximately constant. Therefore, the assumption of homoscedasticity seems to hold.

If this had not been the case, appropriate remedies might include:

- Applying a logarithmic or square root transformation to the dependent variable.
- Using weighted least squares or robust standard errors.
- Re-specifying the model with polynomial or interaction terms to address non-linearity.

2. Independence of Errors The Durbin-Watson statistic is 1.82, which is close to the ideal value of 2.0. This indicates little to no autocorrelation in the residuals.

Durbin-Watson Test for Autocorrelation

After fitting a multiple regression model, one critical assumption to test is the **independence of residuals**. If residuals are correlated, especially in time series or ordered data, this violates the regression assumption of independence and can lead to invalid inference.

The **Durbin-Watson (DW) test** is used to detect the presence of first-order autocorrelation in the residuals (i.e., the relationship between consecutive residuals).

Step-by-Step Instructions

1. **Fit the Regression Model.** Use Excel, R, SPSS, or any other tool to run your multiple regression model.
2. **Obtain the Residuals.** Extract the residuals e_t from your model output.
3. **Compute the DW Statistic:**

$$DW = \frac{\sum_{t=2}^{n}(e_t - e_{t-1})^2}{\sum_{t=1}^{n} e_t^2}$$

   Where:

   - e_t is the residual at time t,
   - n is the number of observations.

4. **Interpret the DW Value:**
    - DW approx. 2 $\rightarrow$ No autocorrelation
    - DW $< 2 \rightarrow$ Positive autocorrelation
    - DW $> 2 \rightarrow$ Negative autocorrelation

    Use published critical values or software to determine significance. For example, in R:

    ```
    install.packages("lmtest")
    library(lmtest)
    dwtest(model)
    ```

    Or in Excel, compute manually:
    - Column A: residuals e_t
    - Column B: $e_t - e_{t-1}$
    - Column C: Square of values in A and B
    - Use SUM formula to complete the DW equation.

Example

Suppose we have the following residuals from our regression model (in time order):

$$\{1.2,\ 0.9,\ -0.3,\ -1.1,\ -0.8,\ 0.5\}$$

- First, compute the numerator:

$$\sum_{t=2}^{6}(e_t - e_{t-1})^2 =$$

$$\begin{aligned} &(0.9 - 1.2)^2 \\ &+ (-0.3 - 0.9)^2 \\ &+ (-1.1 - (-0.3))^2 \\ &+ (-0.8 - (-1.1))^2 \\ &+ (0.5 - (-0.8))^2 \end{aligned}$$

$$= (-0.3)^2 + (-1.2)^2 + (-0.8)^2 + (0.3)^2 + (1.3)^2$$

$$= 0.09 + 1.44 + 0.64 + 0.09 + 1.69 = 3.95$$

- Then compute the denominator:

$$\sum_{t=1}^{6} e_t^2 = 1.44+0.81+0.09+1.21+0.64+0.25 = 4.44$$

- Therefore:

$$DW = \frac{3.95}{4.44} \approx 0.89$$

Interpretation

Since $DW \approx 0.89$, which is significantly less than 2, we suspect **positive autocorrelation** in the residuals.

This violates the independence assumption.

Remedies if Autocorrelation is Present

- Consider including lagged terms or using a time series model (e.g., ARIMA).
- Apply the **Cochrane-Orcutt procedure** or **Prais-Winsten transformation**.
- For panel data, use **Generalized Least Squares (GLS)** or **cluster-robust standard errors**.

3. Homoscedasticity To evaluate whether the assumption of constant variance (homoscedasticity) holds in our multiple regression model, we use the **Breusch-Pagan test**. This test checks for heteroscedasticity by examining whether the variance of the residuals depends on the fitted values (or predictor variables).

- **Null Hypothesis:** The variance of the errors is constant (homoscedasticity).
- **Alternative Hypothesis:** The variance of the errors is not constant (heteroscedasticity).

How to perform the test in R:

1. First, fit your multiple regression model:

```
model <- lm(y ~ X1 + X2 + X3, data = dataset)
```

2. Then, load the `lmtest` package:

```
library(lmtest)
```

3. Finally, run the Breusch-Pagan test:

```
bptest(model)
```

Example Output:

```
studentized Breusch-Pagan test


data:  model
BP = 3.42, df = 3, p-value = 0.3301
```

In our case, we obtained a p-value of 0.41. Since this is greater than the standard significance level of $\alpha = 0.05$, we **fail to reject the null hypothesis**. This means there is no significant evidence of heteroscedasticity, and we can assume that the residuals have approximately constant variance.

If the p-value had been less than 0.05, we would have taken steps to correct for heteroscedasticity, such as:

- Applying a **logarithmic transformation** to the dependent variable,
- Using **robust standard errors** (e.g., `coeftest(model, vcov = vcovHC(model))`),
- Trying **weighted least squares (WLS)** or
- Considering **generalized linear models (GLM)**.

4. Normality of Residuals To validate the assumption that residuals are normally distributed, we conduct both a formal statistical test and a graphical inspection. One widely used formal test is the **Anderson-Darling test**, which evaluates whether a sample of residuals comes from a normal distribution.

- **Null Hypothesis:** The residuals follow a normal distribution.
- **Alternative Hypothesis:** The residuals do not follow a normal distribution.

How to perform the Anderson-Darling test in R:

1. First, fit your multiple regression model and extract the residuals:

```
model <- lm(y ~ X1 + X2 + X3, data = dataset)
residuals <- resid(model)
```

2. Then, install and load the `nortest` package (if not already installed):

```
install.packages("nortest")  # only if needed
library(nortest)
```

3. Finally, run the Anderson-Darling test:

```
ad.test(residuals)
```

Example Output:

```
    Anderson-Darling normality test

data:  residuals
A = 0.3825, p-value = 0.389
```

In our case, the p-value is 0.389, which is greater than the standard significance level of $\alpha = 0.05$. Therefore, we **fail to reject the null hypothesis**, concluding that the residuals are approximately normally distributed.

Visual Confirmation: Q-Q Plot

A Q-Q (quantile-quantile) plot is also a valuable tool for visually inspecting normality. In R, this can be produced using:

```
qqnorm(residuals)
qqline(residuals)
```

If the plotted points fall roughly along the reference line, the residuals are reasonably normal.

If the Anderson-Darling Test Indicates Non-Normality:

- Consider applying a **transformation** (e.g., logarithmic or square root) to the dependent variable.
- Use **robust regression techniques** that are less sensitive to non-normality.
- Evaluate the presence of **outliers** that may be distorting the residual distribution.

5. Multicollinearity Multicollinearity occurs when two or more independent variables in a multiple regression model are highly correlated, which can inflate the variance of the estimated regression coefficients and reduce the model's reliability. A standard diagnostic tool for assessing multicollinearity is the **Variance Inflation Factor (VIF)**.

Definition of VIF:

The VIF for a predictor X_j is calculated as:

$$\text{VIF}_j = \frac{1}{1 - R_j^2}$$

where R_j^2 is the coefficient of determination from a regression of X_j on all the other independent variables. A high R_j^2 (close to 1) indicates that X_j is highly predictable by the other variables, and thus the VIF will be large.

Interpretation Guidelines:

- $\text{VIF} = 1$: No multicollinearity.
- $1 < \text{VIF} \leq 5$: Moderate multicollinearity (usually acceptable).
- $\text{VIF} > 5$: Potentially problematic multicollinearity.
- $\text{VIF} > 10$: Severe multicollinearity, likely distorting regression estimates.

How to Calculate VIFs in R:

To compute VIFs in R, follow these steps:

1. Load your data and fit the multiple regression model:

```
model <- lm(
  Y ~ X1 + X2 + X3,
  data = mydata
)
```

2. Install and load the `car` package (if not already installed):

```
install.packages("car")
library(car)
```

3. Use the `vif()` function to compute the VIFs:

```
vif(model)
```

Example Output:

```
  X1      X2      X3
5.23    6.91    1.34
```

Interpretation:

- $\text{VIF}_{X_1} = 5.23$ and $\text{VIF}_{X_2} = 6.91$: These values exceed the threshold of 5, suggesting moderate-to-high multicollinearity between X_1 and X_2.

- $\text{VIF}_{X_3} = 1.34$: This indicates no significant multicollinearity involving X_3.

Remedies for Multicollinearity:

- **Remove one of the collinear variables**: If X_1 and X_2 are highly correlated and conceptually redundant, consider dropping one of them.
- **Combine variables**: Create an index or composite variable (e.g., average or principal component).
- **Use dimensionality reduction techniques**: Apply *Principal Component Regression (PCR)* or *Partial Least Squares Regression.*
- **Apply regularization methods**: Ridge Regression (L2 penalty) or Lasso (L1 penalty) can stabilize estimates by shrinking coefficients.

In our model, VIFs indicate notable collinearity between X_1 and X_2. We proceeded by examining the correlation matrix and scatter plots, confirming the redundancy. While multicollinearity does not violate regression assumptions per se, it can destabilize coefficient estimates, so careful interpretation and remedial action may be warranted depending on the modeling goals.

Corrective Strategies for Violated Assumptions

If any of the model assumptions are significantly violated, we consider the following remedies:

- **Non-Linearity:** Introduce polynomial terms such as X_1^2, or use transformations (e.g., $\log X$, $\sqrt{X}$).
- **Multicollinearity:** Use Principal Component Regression (PCR), Ridge Regression, or drop redundant predictors.
- **Heteroscedasticity:** Apply log-transformation or use robust standard errors.
- **Non-Normality:** Apply a Box-Cox transformation to the response variable or switch to Generalized Linear Models (GLMs).

Conclusion

This example demonstrates how to perform a thorough multiple regression analysis. Assumption testing ensures the validity of conclusions. Where assumptions are violated, alternative methods exist to improve model robustness and interpretability.

Chapter 15

Nonlinear Regression

15.1 Nonlinearity

Not all relationships between variables follow a straight-line trajectory. While linear regression provides a powerful tool for understanding simple associations, many natural and social phenomena unfold in ways that defy linear explanation. In real-world data, especially from domains like biology, economics, medicine, and environmental science, relationships often demonstrate curves, thresholds, plateaus, saturation effects, or even chaotic dynamics. This necessitates the use of nonlinear regression techniques, which allow us to model relationships that change direction, accelerate, or decelerate as inputs vary.

Nonlinear regression techniques provide a broader set of

tools to capture such complexities. Unlike linear models, where the impact of each predictor is constant across its range, nonlinear models permit the effect of predictors to vary based on their values. This enhanced flexibility is essential when modeling situations such as population growth (which may level off), the spread of diseases (which may initially accelerate and then stabilize), or consumer behavior (which may follow diminishing marginal returns).

A nonlinear model is generally expressed as:

$$y = f(x_1, x_2, \ldots, x_k; \theta_1, \theta_2, \ldots, \theta_p) + \varepsilon$$

where f is a nonlinear function involving the predictors x_i and parameters θ_j, and ε is the error term. The goal is to estimate the parameters θ so that the function best fits the observed data.

Popular types of nonlinear regression models include:

- **Polynomial Regression:** Models nonlinear trends by adding polynomial terms, such as x^2, x^3, etc. This approach is still linear in parameters but nonlinear in the variables. Useful for capturing U-shaped or inverted U-shaped relationships.
- **Exponential Regression:** Often used to describe rapid growth or decay, such as radioactive decay, compound interest, or infection rates. These models take the form $y = ae^{bx}$, where a and b are constants.
- **Logistic and Sigmoidal Models:** Capture

S-shaped curves and are particularly useful in modeling probability outcomes, such as the likelihood of success/failure, disease spread, or adoption of technologies. The logistic function is bounded between 0 and 1, making it ideal for binary responses.

- **Spline Regression:** Fits piecewise polynomials to subsets of the data, joined at knot points. This technique is very effective for modeling data where the relationship changes form over different ranges. Natural splines and B-splines are commonly used variants.
- **Kernel Regression:** A non-parametric technique that uses weighted averages of nearby observations to estimate the conditional expectation of y given x. It does not assume any specific functional form and is suitable for smooth trend discovery in high-noise settings.
- **Neural Networks:** Inspired by biological neurons, these models use layers of computational nodes to learn patterns from data. Deep neural networks (DNNs) can capture highly nonlinear relationships by composing multiple layers of nonlinear transformations. They are widely used in image recognition, natural language processing, and forecasting.

These methods allow for great adaptability in modeling complex, real-world processes. However, they also come with challenges: overfitting, interpretability, and computational demands increase as model complexity grows. Therefore, care must be taken when selecting a nonlinear model to ensure it reflects the underlying reality of the data while maintaining robustness and clarity.

Moreover, while nonlinear regression increases modeling flexibility, it often requires iterative estimation techniques, such as Gauss-Newton or gradient descent algorithms. Unlike linear regression, closed-form solutions typically do not exist for nonlinear models, so numerical optimization methods must be used.

Ultimately, embracing nonlinear regression is a powerful next step in statistical modeling. It equips analysts and researchers with the ability to move beyond the straight line, discovering deeper truths and more accurate representations of data that would remain hidden under rigid linear assumptions.

15.2 Transforming Models

Linear regression, though powerful in its simplicity, often falls short in modeling the true structure of complex data. In practice, relationships between variables may not conform neatly to a straight-line equation. To address this, one powerful approach is to transform a linear model into a form that captures nonlinear dynamics while preserving the interpretability and structure of linear regression. These transformations can enhance the model's flexibility and adherence to assumptions, enabling better predictive performance and clearer interpretation.

Transformations are used to reshape the predictor variables or the response variable in a way that reveals hidden structure or corrects for issues such as non-linearity, non-constant variance, or skewed distributions. The transformed models are still estimated using familiar linear regression methods, but the underlying functional relationship is nonlinear.

- **Log Transformations:** Applying the natural logarithm to either the response variable (y) or a predictor (x) helps linearize exponential relationships, stabilize variance (addressing heteroscedasticity), and reduce right-skewness in the data. For instance, $\log(y) = \beta_0 + \beta_1 x$ can be used when y grows exponentially.

- **Polynomial Terms:** Including powers of a predictor such as $x^2, x^3, \ldots$ introduces curvature to the model. This is particularly useful when a scatterplot shows a bowl or hill shape. Even though the relationship is nonlinear in x, it remains linear in the coefficients, allowing OLS estimation.

- **Square Root or Inverse Transformations:** These are effective when the response variable exhibits diminishing returns. For example, if increases in x lead to progressively smaller increases in y, then a square root or inverse function can linearize the relationship.

- **Box-Cox Transformation:** This family of power transformations identifies an optimal transformation from a continuum of options (e.g., log, square root, reciprocal) that improves normality and variance stabilization. The transformation is defined as:

$$y^{(\lambda)} = \begin{cases} \frac{y^\lambda - 1}{\lambda}, & \lambda \neq 0 \\ \log(y), & \lambda = 0 \end{cases}$$

It is data-driven and can be determined using maximum likelihood estimation.

- **Principal Component Regression (PCR):** When predictors are highly correlated, PCR reduces them to a smaller set of uncorrelated principal components. These new variables retain most of the variance of the original predictors and can help capture nonlinear interaction effects.

- **Ridge Regression:** This regularization technique addresses multicollinearity by adding a penalty term to the regression objective function. Although ridge does not explicitly transform variables, it allows for stable estimation in complex or high-dimensional settings where nonlinear dynamics often arise.

These techniques often operate within the framework of generalized linear models (GLMs), extending the reach of OLS by changing the link function or the assumed distribution of the response variable. For example, a log link with a Poisson distribution is commonly used in count data modeling.

The selection of the appropriate transformation requires domain knowledge, exploratory data analysis, and diagnostic checking. Plots such as residual plots, partial regression plots, and scatterplots of transformed variables are indispensable tools. Furthermore, automated methods such as the Box-Tidwell test and model selection criteria (AIC, BIC) assist in evaluating transformation adequacy.

Ultimately, transforming linear models into nonlinear ones allows statisticians and analysts to strike a balance between interpretability and flexibility. These approaches offer a practical path to uncovering complex relationships without fully departing from the familiar and robust foundation of linear modeling.

Transforming Linear Models into Nonlinear Ones

Linear regression, though powerful in its simplicity, often falls short in modeling the true structure of complex data. In practice, relationships between variables may not conform neatly to a straight-line equation. To address this, one powerful approach is to transform a linear model into a form that captures nonlinear dynamics while preserving the interpretability and structure of linear regression. These transformations can enhance the model's flexibility and adherence to assumptions, enabling better predictive performance and clearer interpretation.

Transformations are used to reshape the predictor variables or the response variable in a way that reveals hidden structure or corrects for issues such as non-linearity, non-constant variance, or skewed distributions. The transformed models are still estimated using familiar linear regression methods, but the underlying functional relationship is nonlinear.

- **Log Transformations:** Applying the natural logarithm to either the response variable (y) or a predictor (x) helps linearize exponential relationships, stabilize variance (addressing heteroscedasticity), and reduce right-skewness in the data. For instance, $\log(y) = \beta_0 + \beta_1 x$ can be used when y grows exponentially. Log-log models ($\log(y) = \beta_0 + \beta_1 \log(x)$) are often used when both variables grow multiplicatively.

- **Polynomial Terms:** Including powers of a predictor such as $x^2, x^3, \ldots$ introduces curvature to the model. This is particularly useful when a scatterplot shows a bowl or hill shape. Even though the relationship is nonlinear in x, it remains linear in the coefficients, allowing OLS estimation. Interaction terms like $x_1 x_2$ can also model combined effects of variables.

- **Square Root or Inverse Transformations:** These are effective when the response variable exhibits diminishing returns. For example, if increases in x lead to progressively smaller increases in y, then a square root or inverse function can linearize the relationship. Common forms include $\sqrt{x}$ and $1/x$.

- **Box-Cox Transformation:** This family of power transformations identifies an optimal transformation from a continuum of options (e.g., log, square root, reciprocal) that improves normality and variance stabilization. The transformation is defined as:

$$y^{(\lambda)} = \begin{cases} \frac{y^\lambda - 1}{\lambda}, & \lambda \neq 0 \\ \log(y), & \lambda = 0 \end{cases}$$

It is data-driven and can be determined using maximum likelihood estimation. The Box-Cox procedure is supported in most statistical software (e.g., R: 'boxcox()' function in the 'MASS' package).

- **Principal Component Regression (PCR):** When predictors are highly correlated, PCR reduces them to a smaller set of uncorrelated principal components. These new variables retain most of the variance of the original predictors and can help capture nonlinear interaction effects. PCR is especially helpful in high-dimensional spaces or when predictors outnumber observations.

- **Ridge Regression:** This regularization technique addresses multicollinearity by adding a penalty term to the regression objective function. Although ridge does not explicitly transform variables, it allows for stable estimation in complex or high-dimensional settings where nonlinear dynamics often arise. The ridge penalty $\lambda \sum \beta_j^2$ shrinks coefficients, discouraging extreme values.

- **Generalized Linear Models (GLMs):** GLMs extend linear models by allowing for response variables that follow distributions from the exponential family (e.g., binomial, Poisson). By specifying a link function (e.g., logit, log), GLMs can model a variety of nonlinear phenomena while still using linear predictors.
- **Spline Transformations:** When there are suspected changes in slope or non-continuous curvature, spline methods (e.g., natural splines, B-splines) fit piecewise polynomials to the data. This can accommodate local nonlinear patterns while controlling for overfitting.
- **Box-Tidwell Transformation:** This test evaluates whether a transformation of the predictor variables improves the linear fit of a regression model. It's an extension of the Box-Cox method but is applied to predictors instead of the response.

These techniques often operate within the framework of generalized linear models (GLMs), extending the reach of OLS by changing the link function or the assumed distribution of the response variable. For example, a log link with a Poisson distribution is commonly used in count data modeling.

The selection of the appropriate transformation requires domain knowledge, exploratory data analysis, and diagnostic checking. Plots such as residual plots, partial regression plots, and scatterplots of transformed variables are indispensable tools. Furthermore, automated methods such as the Box-Tidwell test and model selection criteria (AIC, BIC) assist in evaluating transformation adequacy.

Ultimately, transforming linear models into nonlinear ones allows statisticians and analysts to strike a balance between interpretability and flexibility. These approaches offer a practical path to uncovering complex relationships without fully departing from the familiar and robust foundation of linear modeling.

15.3 Dimension Reduction

Sometimes models suffer from too many predictors, multicollinearity, or irrelevant variables that introduce noise into the estimation process. These challenges often arise when datasets are high-dimensional, when predictors are highly correlated, or when sample sizes are small relative to the number of variables. In such cases, techniques that reduce dimensionality or regularize estimates are essential for producing interpretable and stable models. Two widely-used approaches are Principal Component Regression (PCR) and Ridge Regression.

- **Principal Component Regression (PCR):** PCR is a two-stage procedure that begins with Principal Component Analysis (PCA) to transform the original predictors into a new set of orthogonal components (principal components). These components capture the directions of maximum variance in the data. Rather than regressing the response variable on the raw predictors, PCR uses a selected number of these components as inputs. This reduces multicollinearity and dimensionality while retaining most of the information present in the original variables. The analyst must determine how many components to retain—usually based on a scree plot or cumulative variance explained (e.g., retaining components that explain 90-95% of total variance).

- **Ridge Regression:** Ridge Regression modifies the OLS estimation process by adding a penalty term to the loss function. The goal is to minimize the residual sum of squares while simultaneously shrinking the coefficients:

$$\text{RSS}_{\text{ridge}} = \sum_{i=1}^{n}(y_i - \hat{y}_i)^2 + \lambda \sum_{j=1}^{p} \beta_j^2$$

  The tuning parameter λ controls the strength of the penalty. When $\lambda = 0$, the model reduces to standard OLS. As λ increases, coefficients shrink toward zero. This reduces model complexity, mitigates multicollinearity, and guards against overfitting—especially valuable in high-dimensional datasets. Ridge regression can be implemented in R using the 'glmnet' package.

- **Elastic Net and Lasso (optional extensions):** While not the focus here, it is worth noting that related regularization techniques include the Lasso (L_1 penalty) and Elastic Net (a combination of L_1 and L_2). Lasso tends to produce sparse models with variable selection, while Ridge retains all variables but shrinks their impact.

- **When to Use:** These methods are ideal when:
  - The predictor variables exhibit strong correlation.
  - The number of predictors is large relative to the sample size.
  - The goal is prediction and model stability, not just explanation.
- **Interpretation Caveats:** While powerful, these techniques can obscure the interpretability of coefficients. PCR components are combinations of variables, not directly observable traits. Ridge estimates are biased but often have lower mean squared error. Thus, communication of results should include a discussion of trade-offs.

In summary, dimension reduction and regularization provide essential tools for modern statistical modeling. They help prevent overfitting, reduce noise, and offer solutions when classic assumptions break down. When combined with diagnostics and validation, they enhance the robustness and generalizability of regression models.

A Statistician's Discipline

Regardless of whether the model is linear or nonlinear, all regression models rest on core assumptions. These assumptions are the bedrock upon which reliable inference, predictive validity, and scientific credibility rest. If any of these assumptions are violated, the results of the regression model may be misleading or entirely invalid. Thus, the disciplined statistician or data scientist must always pause to verify them systematically.

Core Assumptions

- **Linearity:** The relationship between predictors and the outcome should be linear in the parameters, or transformed to be so. This can be assessed visually using scatterplots and partial regression plots.
- **Independence of Residuals:** The residuals (errors) should be independent of each other, meaning there should be no pattern or autocorrelation over time or space. This is crucial in time series and spatial data analysis.
- **Homoscedasticity (Constant Variance):** The variance of residuals should be roughly constant across levels of the independent variables. Heteroscedasticity may inflate Type I errors.
- **Normality of Residuals:** The residuals should be approximately normally distributed, particularly for inference. This is less crucial for prediction but essential for hypothesis testing.
- **No or Low Multicollinearity:** Independent variables should not be too strongly correlated with each other. Multicollinearity can lead to unstable estimates and inflated standard errors.

Tools and Techniques for Testing Assumptions

- **Residual Plots:** Plot residuals against fitted values and individual predictors to check for non-linearity and heteroscedasticity.
- **Durbin-Watson Test:** Used to detect autocorrelation in residuals, especially in time series data. A value near 2 indicates independence.
- **Breusch-Pagan Test:** This tests for heteroscedasticity. A non-significant p-value supports the assumption of constant variance.
- **Anderson-Darling Test or Shapiro-Wilk Test:** These are used to test whether residuals are normally distributed.
- **Variance Inflation Factors (VIF):** These quantify multicollinearity. A VIF above 5–10 suggests a problematic level of collinearity.

Instructional Tips for Novices

1. **Plot First.** Always start with graphical diagnostics. They are intuitive and often more revealing than raw numbers.

2. **Use Built-In Functions.** In R, for instance:

```
lm_model <- lm(y ~ x1 + x2 + x3,
data = dataset)
```

```
plot(lm_model)
library(lmtest) #residuals v fitted
bptest(lm_model) # Breusch-Pagan test
library(car)
vif(lm_model) # VIF scores
library(nortest)
ad.test(residuals(lm_model)) #ad test
```

3. **Interpret Carefully.** No single test gives the full picture. Look for convergence among multiple diagnostics.

4. **Transform or Redesign When Needed.** If assumptions fail, consider data transformations (e.g., log, square root), removing outliers, or using robust regression techniques.

5. **Context Matters.** Consider the nature of your data—e.g., independence may not be assumed in clustered or longitudinal data. In such cases, mixed models or generalized estimating equations may be more appropriate.

Discipline and Rigor

Statistical modeling is not just about running code—it's about adopting a disciplined mindset. Great data scientists are not just coders but methodical problem solvers who interrogate their models. Each regression output must be scrutinized. If the assumptions hold, we can be more confident that our estimates, p-values, and predictions are meaningful. If not, we revise, refine, and revalidate.

By consistently checking assumptions and understanding their implications, we elevate the practice of regression from mere computation to robust, defensible scientific analysis.

15.4 Parsimony

When building models, it's crucial to avoid two major pitfalls that can compromise the quality and interpretability of your statistical analysis:

- **Too Few Variables:** Omitting relevant predictors introduces bias, distorts relationships among included variables, and may leave confounding factors unaddressed. This is known as underfitting. The model may miss important patterns and yield poor predictive accuracy.
- **Too Many Variables:** Including too many predictors, especially those that contribute little unique information, inflates the variance of coefficient estimates and increases the risk of overfitting. This complicates interpretation, weakens generalizability, and can cause the model to perform poorly on new data.

Choosing an appropriate model involves balancing complexity with simplicity—a principle known as **parsimony**. A parsimonious model captures the essential structure of the data with the fewest number of predictors necessary to retain predictive accuracy and explanatory power.

Model Selection Strategies

To achieve parsimony, several techniques can be used:

- **Forward Selection:** Start with no variables and add predictors one at a time based on which improves the model most (e.g., lowest AIC or highest adjusted R^2).
- **Backward Elimination:** Start with all predictors and remove the least significant one at each step.
- **Stepwise Regression:** A combination of forward and backward methods, evaluating variables for inclusion and exclusion iteratively.
- **Information Criteria:** The Akaike Information Criterion (AIC) and Bayesian Information Criterion (BIC) balance model fit with complexity. Lower values suggest better models.
- **Cross-Validation:** Helps evaluate model generalizability by testing performance on unseen data.

The Curse of Dimensionality and the Need for Caution

As the number of predictors grows, the data become sparse in high-dimensional space. This is referred to as the "curse of dimensionality." It can lead to models that are overly tailored to the training data but fail to generalize. High-dimensional settings also increase the likelihood of multicollinearity, which makes coefficient estimates unstable.

Parsimony in Practice

In real-world data science, parsimony is not just a theoretical ideal—it is a practical necessity. Models must be:

- **Interpretable:** Stakeholders must understand the model's logic.
- **Predictively Strong:** Must generalize well to new observations.
- **Computationally Efficient:** Especially when deployed in real-time systems.

Avoiding overcomplexity doesn't mean oversimplifying. It means resisting the temptation to add predictors without a clear rationale, and instead selecting features informed by domain knowledge, exploratory data analysis, and rigorous statistical testing.

Conclusion

Parsimony is about modeling with discipline—favoring elegance and efficiency without sacrificing accuracy. The most useful model is not the one with the most variables, but the one that best balances fidelity to the data with clarity and generalizability. In the hands of a thoughtful analyst, parsimony becomes not just a principle, but a powerful tool for scientific insight.

A well-constructed parsimonious model does not merely summarize a dataset—it reveals underlying structures, highlights meaningful trends, and resists being misled by noise. Such models stand up better under cross-validation and are more robust in dynamic environments where future data may shift subtly.

Ultimately, parsimony teaches us to respect the complexity of the real world while striving for clarity in our explanations. It pushes us to validate each additional variable, to test rigorously, and to refine our models not by brute force but by the wisdom of thoughtful inclusion. In the quest for knowledge and application, it reminds us that precision and restraint often yield the most enduring insights.

15.5 Conclusion

With the foundational principles of regression in hand—checking assumptions, interpreting diagnostics, and selecting models wisely—we are empowered to tackle even the most complex modeling problems. Whether applying polynomial fits to economic data, using logistic models for classification, or experimenting with neural networks, our training ensures that we never lose sight of statistical validity. Through diligent application and model testing, we strive to make inferences that are not only meaningful but statistically sound and scientifically defensible.

The Path from Theory to Mastery

In practice, the art of modeling requires a delicate balance between the theoretical rigor we've been trained in and the flexibility demanded by real-world data. No dataset arrives neatly packaged with clean assumptions. Instead, the data scientist must navigate missing values, detect outliers, identify nonlinearities, and apply thoughtful transformations. This journey from messy data to reliable insight is where training meets wisdom.

Our regression toolkit—OLS, GLM, nonlinear techniques, and regularization methods—is more than a menu of options. It represents a system of thinking. Every time we decide to center or standardize variables, check variance inflation factors, inspect residual plots, or run diagnostics such as the Breusch-Pagan or Durbin-Watson tests, we are honoring the core principles of sound inference.

Modeling in an Era of Big Data

As we enter an era of increasingly large and complex datasets, the demand for thoughtful and defensible modeling has never been higher. Machine learning models can fit data with astonishing accuracy, but their lack of interpretability often leads to questions about reliability and transparency. That's where statistical regression retains its vital role—offering models that explain as well as predict.

Even in high-dimensional spaces, the foundational tools we've learned help us avoid the perils of spurious correlations and unstable estimations. Dimension reduction, parsimony, regularization, and cross-validation remain essential no matter how powerful our computational tools become.

The Ethical Imperative

Finally, responsible modeling is not merely a technical obligation—it is an ethical one. Our conclusions influence policy decisions, business strategies, and social understanding. Inaccurate models may mislead, while well-tested models have the potential to illuminate truth. Thus, the responsibility on the shoulders of statisticians and data scientists is immense.

We must always be asking:

- Are my assumptions valid?
- Are my conclusions supported by evidence?
- Have I considered bias, confounding, and omitted variable risks?
- Does this model clarify, or does it obscure?

Final Reflection

The study of regression is not just a methodological milestone; it is a gateway to deeper statistical insight. It teaches us to see relationships, to question patterns, and to test hypotheses with care. As we advance, these principles become the scaffolding upon which we build more complex models and systems.

Training meets practice not in perfection, but in the discipline to ask the right questions, use the right tools, and check our assumptions—always.

We are not just data analysts; we are translators of evidence into understanding. And in that mission, statistical rigor remains our compass.

Chapter 16

Advanced Methods

16.1 Introduction

While traditional parametric methods assume specific data distributions and model forms, many real-world situations demand more flexible, robust, or specialized techniques. This chapter explores powerful tools used when parametric assumptions are not met, when data are observed across time, or when businesses require operational optimization and predictive planning.

16.2 Nonparametric Methods

Nonparametric methods do not assume a particular population distribution, making them useful for ordinal data, small samples, or unknown distributions.

16.2.1 Sign Test

- **Hypotheses:**
  - Null Hypothesis (H_0): The median difference between the paired observations is zero.
  - Alternative Hypothesis (H_1): The median difference is not zero (or is greater/less than zero, in one-tailed tests).
- **Assumptions:**
  - The data are paired and randomly selected.
  - Each pair is independent of other pairs.
  - The measurement scale is at least ordinal.
- **Procedure:**
  1. Compute the differences between each pair.
  2. Discard pairs where the difference is zero.
  3. Count the number of positive and negative differences.
  4. Use the binomial distribution to test whether the observed number of positive (or negative) signs differs significantly from what would be expected by chance.
- **Test Statistic:** The number of positive signs (or the smaller of the number of positive and negative signs, in a two-tailed test).
- **P-value:** Computed using the binomial distribution based on the number of non-zero

pairs and observed signs.

- **Use Case:** The sign test is useful when comparing two treatments or conditions applied to the same subjects. For example, a hospital might test whether a new surgical technique leads to faster recovery times compared to the standard approach. If each patient receives both treatments (e.g., in different phases or for different conditions), the recovery time differences can be compared using the sign test.
- **Advantages:**
  - No need to assume normality.
  - Simple to compute and interpret.
  - Robust to outliers and skewed data.
- **Limitations:**
  - Less powerful than parametric counterparts like the paired t-test when assumptions for parametric tests are met.
  - Ignores the magnitude of the differences.

16.2.2 Moving Averages

Moving averages are fundamental tools in time series analysis, smoothing fluctuations to reveal underlying patterns such as trends and seasonality. These methods form the backbone of forecasting techniques in economics, operations, inventory management, and financial market analysis.

- **Simple Moving Average (SMA):** This method computes the average of a fixed number of the most recent observations. It is best used when data exhibit a stable pattern without trend or seasonality.
  - *Formula:*

$$\text{SMA}_t = \frac{1}{n} \sum_{i=0}^{n-1} y_{t-i}$$

  - **Use Case:** Inventory reordering policies, smoothing stock prices.
- **Weighted Moving Average (WMA):** Unlike SMA, this method gives more importance to recent data points. Weights typically decrease linearly or exponentially as the observations get older.
  - *Formula:*

$$\text{WMA}_t = \sum_{i=0}^{n-1} w_i y_{t-i} \text{ where } \sum w_i = 1$$

  - **Use Case:** Emphasizing the latest sales figures in fast-changing industries.
- **Exponential Smoothing:** Applies exponentially decreasing weights, giving highest priority to the most recent observation. The smoothing constant $\alpha \in [0, 1]$ determines how quickly the weights

decay.

  - *Formula:*

$$S_t = \alpha y_t + (1 - \alpha) S_{t-1}$$

  - **Use Case:** Short-term demand forecasting, weather prediction.

- **Holt-Winters Method:** An advanced form of exponential smoothing that adjusts for both trend and seasonality. It comes in additive and multiplicative versions depending on the nature of seasonal effects.
  - **Components:** Level, trend, and seasonal indices updated recursively.
  - **Use Case:** Forecasting retail sales, airline passengers, electricity usage with pronounced seasonal patterns.
- **Advantages:**
  - Easy to implement and interpret.
  - Excellent for short-term forecasts.
  - Useful when data lack clear structure.
- **Limitations:**
  - SMA and WMA lag behind turning points.
  - Cannot forecast future turning points without external input.

  - Performance deteriorates with highly volatile or non-stationary data.

- **Example in R:**

```
library(TTR)
data <- c(12, 15, 14,
18, 20, 19, 22)
SMA(data, n = 3)  # Simple MA
EMA(data, n = 3)  # Exponential MA
```

- **Example in Python:**

```
import pandas as pd
data = pd.Series([12, 15, 14,
 18, 20, 19, 22])
data.rolling(window=3).mean()# SMA
data.ewm(span=3, adjust=False).mean()# EMA
```

16.2.3 Wilcoxon Test

The Wilcoxon Signed-Rank Test is a widely used nonparametric alternative to the paired-samples t-test. Unlike the sign test, which considers only the direction of differences, the Wilcoxon test takes into account both the direction and the magnitude of differences between paired observations. This allows it to detect more subtle shifts in central tendency, making it a more powerful option when the assumption of normality in paired differences is violated.

It is particularly valuable when analyzing data that are ordinal, skewed, or influenced by outliers, but still maintain a consistent ranking or ordering across pairs. The test is based on ranking the absolute differences between pairs, assigning signs based on whether the original differences were positive or negative, and then summing these signed ranks.

- **Hypotheses:**
  - Null Hypothesis (H_0): The median difference between paired observations is zero.
  - Alternative Hypothesis (H_1): The median difference is not zero (two-tailed), or is greater/less than zero (one-tailed).
- **Assumptions:**
  - Paired samples are randomly selected and dependent.
  - Differences between pairs are symmetrically distributed around the median.
  - The measurement scale is at least interval or ordinal.
- **Procedure:**
  1. Calculate the difference between each pair.
  2. Exclude any pairs with a difference of zero.
  3. Rank the absolute values of the non-zero differences.

4. Assign the signs of the original differences to their respective ranks.
5. Compute the test statistic W, the smaller of the sum of positive or negative signed ranks.
6. Use statistical tables or software to determine the p-value based on the distribution of W.

- **Test Statistic:** W, the smaller of the sum of the positive or negative ranks.
- **P-value:** Computed from the distribution of the Wilcoxon statistic under the null hypothesis.
- **Use Case:** The Wilcoxon Signed-Rank Test is appropriate for evaluating changes in patient health scores before and after a treatment, comparing exam scores before and after tutoring, or analyzing satisfaction ratings for two products reviewed by the same set of users.
- **Advantages:**
  - More powerful than the sign test since it incorporates magnitude.
  - Does not assume normality, making it robust for skewed or ordinal data.
  - Applicable to small sample sizes.
- **Limitations:**
  - Less efficient than the paired t-test when data are normally distributed.

  - Requires symmetrical distribution of differences for valid inference.
  - Cannot handle ties or zero differences without adjustments.

- **Software Support:** Widely available in statistical packages like R (`wilcox.test()`), Python (`scipy.stats.wilcoxon`), SPSS, SAS, and Excel add-ins.

- **Example in R:**

```
before <- c(70, 65, 80, 90, 85)
after  <- c(75, 70, 78, 92, 83)
wilcox.test(before, after, paired = TRUE)
```

  This test would compare the paired scores and return a p-value indicating whether the median difference is significantly different from zero.

The Wilcoxon Signed-Rank Test strikes a balance between robustness and sensitivity, offering a nuanced approach to analyzing paired data when parametric assumptions are not met.

16.2.4 Mann-Whitney

The Mann-Whitney U Test, also known as the Wilcoxon Rank-Sum Test, is a nonparametric method used to compare differences between two independent groups when the assumptions of a traditional independent-samples t-test are not met. This test evaluates whether the distributions of two groups differ significantly, without assuming a normal distribution of the data. It is particularly useful for ordinal data, or for continuous data that exhibit skewness or contain outliers.

Unlike the t-test, which compares means, the Mann-Whitney U test compares the median ranks of two groups. It is based on the relative ranking of the data from both groups combined, making it robust against non-normality and heteroscedasticity.

- **Hypotheses:**
  - Null Hypothesis (H_0): The distributions of both groups are equal.
  - Alternative Hypothesis (H_1): One distribution is stochastically greater than the other.
- **Assumptions:**
  - Observations are independent between groups.
  - The dependent variable is ordinal or continuous.
  - The shapes of the distributions are similar.
- **Procedure:**
  1. Combine the data from both groups.
  2. Rank all values from smallest to largest, assigning average ranks in the case of ties.

3. Calculate the sum of ranks for each group.
4. Compute the U statistic using:

$$U = n_1 n_2 + \frac{n_1(n_1+1)}{2} - R_1$$

   where n_1 and n_2 are the sample sizes of groups 1 and 2, and R_1 is the sum of the ranks for group 1.

5. Determine the p-value using the U distribution (or normal approximation if both n_1 and n_2 are large).

- **Test Statistic:** U, the smaller of the two U values computed for each group.
- **P-value:** Derived from the U distribution or normal approximation.
- **Use Case:** Comparing salaries of employees in two company divisions, evaluating treatment vs. control in medical trials with non-normal outcome distributions, or assessing user ratings across two different app versions.
- **Advantages:**
  - Does not require normal distribution.
  - Can handle ordinal and skewed data.
  - Robust to outliers.
- **Limitations:**

  - Less powerful than the t-test under normality.
  - Assumes similar distribution shapes; otherwise, median comparisons may be misleading.
  - Sensitive to unequal variances when distributions differ.

- **Software Support:** Available in R (`wilcox.test()` with `paired = FALSE`), Python (`scipy.stats.mannwhitneyu`), SPSS, SAS, and other statistical tools.
- **Example in R:**

```
group1 <- c(45, 50, 55, 60, 65)
group2 <- c(52, 58, 63, 67, 70)
wilcox.test(group1, group2,
paired = FALSE)
```

  This command performs the Mann-Whitney U test to compare whether the two independent samples come from the same distribution.

The Mann-Whitney U test is a flexible and robust tool that allows for meaningful comparison of independent samples in the absence of normality, making it invaluable in real-world applications where ideal assumptions rarely hold.

16.2.5 Kruskal-Wallis Test

The Kruskal-Wallis H test is a nonparametric method used to test whether samples originate from the same distribution. It extends the Mann-Whitney U test to more than two independent groups and is often used as a nonparametric alternative to one-way ANOVA when the assumptions of normality or homogeneity of variances are violated.

- **Hypotheses:**
  - Null Hypothesis (H_0): The medians of all groups are equal.
  - Alternative Hypothesis (H_1): At least one group differs from the others.
- **Assumptions:**
  - The samples are independent.
  - The dependent variable is ordinal or continuous.
  - The shape of the distributions should be similar among groups.
- **Procedure:**
  1. Combine all group data into a single list.
  2. Rank all observations from smallest to largest.
  3. Compute the average rank for each group.

4. Use the Kruskal-Wallis H formula:

$$H = \frac{12}{N(N+1)} \sum_{i=1}^{k} \frac{R_i^2}{n_i} - 3(N+1)$$

   where N is the total number of observations, k is the number of groups, R_i is the sum of ranks for group i, and n_i is the sample size for group i.

5. Compare H to the chi-squared distribution with $k-1$ degrees of freedom to obtain a p-value.

- **Use Case:** Evaluating satisfaction scores across several departments in a company, comparing test scores among multiple schools, or measuring effectiveness of multiple treatments in clinical trials.
- **Advantages:**
  - Does not assume normal distribution or equal variances.
  - Applicable to ordinal and non-normally distributed data.
  - Easy to compute using standard statistical software.
- **Limitations:**
  - Does not indicate which groups differ — follow-up tests (e.g., Dunn's test) are required.

  - Assumes similar shaped distributions across groups.

- **Software Support:** Available in R (`kruskal.test()`), Python (`scipy.stats.kruskal`), SPSS, and other platforms.

- **Example in R:**

```
department1 <- c(6, 7, 8)
department2 <- c(5, 6, 7)
department3 <- c(7, 8, 9)
kruskal.test(list(department1,
department2, department3))
```

  This command performs the Kruskal-Wallis test on three department satisfaction scores.

The Kruskal-Wallis test provides a robust method for analyzing group differences when data violate ANOVA assumptions. It is a critical component of nonparametric statistics, offering clarity and insight when parametric tests fall short.

16.2.6 Spearman's Rank

A nonparametric measure of monotonic association between two variables.

- **Use Case:** Assessing correlation between experience rank and satisfaction rank.

16.3 Time Series Analysis

Time series data consist of observations taken sequentially over time. Unlike cross-sectional data, time series analysis captures dependencies across time lags.

16.3.1 Components

- **Trend:** Long-term upward or downward movement.
- **Seasonality:** Regular pattern repeated at fixed intervals.
- **Cyclical Variations:** Long-term fluctuations not tied to calendar.
- **Irregular (Noise):** Random variations.

16.3.2 Stationarity

A stationary time series has constant mean and variance over time. Many models, including ARIMA, require stationarity. Differencing or transformation may be applied to achieve this.

16.3.3 Autocorrelation

- **Autocorrelation Function (ACF):** Measures correlation between time lags.
- **Partial ACF (PACF):** Measures correlation controlling for intermediate lags.

16.4 Forecasting Techniques

16.4.1 Moving Averages

Moving averages are fundamental tools in time series analysis, smoothing fluctuations to reveal underlying patterns such as trends and seasonality. These methods form the backbone of forecasting techniques in economics, operations, inventory management, and financial market analysis.

- **Simple Moving Average (SMA):** This method computes the average of a fixed number of the most recent observations. It is best used when data exhibit a stable pattern without trend or seasonality.
  - *Formula:*

$$\text{SMA}_t = \frac{1}{n} \sum_{i=0}^{n-1} y_{t-i}$$

  - **Use Case:** Inventory reordering policies, smoothing stock prices.
- **Weighted Moving Average (WMA):** Unlike SMA, this method gives more importance to recent data points. Weights typically decrease linearly or exponentially as the observations get older.

  - *Formula:*

$$\text{WMA}_t = \sum_{i=0}^{n-1} w_i y_{t-i} \text{ where } \sum w_i = 1$$

  - **Use Case:** Emphasizing the latest sales figures in fast-changing industries.

- **Exponential Smoothing:** Applies exponentially decreasing weights, giving highest priority to the most recent observation. The smoothing constant $\alpha \in [0, 1]$ determines how quickly the weights decay.

  - *Formula:*

$$S_t = \alpha y_t + (1 - \alpha) S_{t-1}$$

  - **Use Case:** Short-term demand forecasting, weather prediction.

- **Holt-Winters Method:** An advanced form of exponential smoothing that adjusts for both trend and seasonality. It comes in additive and multiplicative versions depending on the nature of seasonal effects.

  - **Components:** Level, trend, and seasonal indices updated recursively.

  - **Use Case:** Forecasting retail sales, airline passengers, electricity usage with pronounced seasonal patterns.

- **Advantages:**
  - Easy to implement and interpret.
  - Excellent for short-term forecasts.
  - Useful when data lack clear structure.
- **Limitations:**
  - SMA and WMA lag behind turning points.
  - Cannot forecast future turning points without external input.
  - Performance deteriorates with highly volatile or non-stationary data.
- **Example in R:**

```
library(TTR)
data <- c(12, 15, 14, 18,
20, 19, 22)
SMA(data, n = 3)  # Simple MA
EMA(data, n = 3)  # Exponential MA
```

- **Example in Python:**

```
import pandas as pd
data = pd.Series([12, 15, 14,
18, 20, 19, 22])
data.rolling(window=3).mean()# SMA
data.ewm(span=3, adjust=False).mean()
```

Moving averages offer a simple yet effective way to understand temporal trends and are essential components of more complex forecasting models. Their use in both operational planning and advanced analytics reflects their enduring utility across industries.

16.4.2 ARIMA Models

The AutoRegressive Integrated Moving Average (ARIMA) model is a powerful and flexible class of models for time series forecasting. It is widely used in economics, finance, supply chain planning, and environmental monitoring. ARIMA combines three core ideas—autoregression (AR), integration (I), and moving average (MA)—into a unified framework that models both trends and patterns in time series data.

- **AR (Autoregressive):** The current value depends on its own past values. Represented as AR(p), where p is the number of lagged observations.
- **I (Integrated):** Differencing is applied to the series to make it stationary (removing trend or seasonality). Represented as I(d), where d is the degree of differencing.
- **MA (Moving Average):** The current value depends on past forecast errors. Represented as MA(q), where q is the size of the moving average window.

Steps to Fit an ARIMA Model:

1. **Visualize the Time Series:** Plot the raw data to identify trends, seasonality, or abrupt shifts.
2. **Check for Stationarity:** Use the Augmented Dickey-Fuller (ADF) test or KPSS test to determine if differencing is required.
3. **Apply Differencing:** Difference the data until the mean and variance stabilize.
4. **Examine ACF and PACF Plots:** Use autocorrelation (ACF) and partial autocorrelation (PACF) plots to estimate values for p and q.
5. **Fit the ARIMA Model:** Use the selected (p,d,q) parameters and evaluate the model using information criteria (AIC, BIC).
6. **Diagnose Residuals:** Residuals should resemble white noise—uncorrelated and normally distributed.
7. **Forecast:** Generate out-of-sample forecasts with prediction intervals.

Benefits:

- Models trends, seasonality, and irregular patterns effectively.
- Produces interpretable coefficients and flexible forecasts.
- Suitable for many real-world applications including demand planning and financial market analysis.

Limitations:

- Requires stationary data, which may not always be achievable.
- Sensitive to the selection of p, d, q parameters.
- Does not handle multiple seasonalities or external regressors unless extended to SARIMA or ARIMAX models.

Example in R:

```
library(forecast)
data <- ts(c(100, 108, 115, 130, 128, 140))
fit <- auto.arima(data)
forecast(fit, h=3)
```

This code uses the `forecast` package to automatically fit the best ARIMA model and forecast the next 3 time periods.

ARIMA remains one of the cornerstone models in time series analysis. By carefully diagnosing the data and following methodical steps, analysts can build robust models that not only describe historical patterns but also project future values with confidence.

16.4.3 SARIMA

Seasonal ARIMA (SARIMA) models extend traditional ARIMA models by explicitly incorporating seasonality into the forecasting process. Denoted as $(p, d, q)(P, D, Q)_s$, SARIMA introduces an additional layer of autoregressive (AR), differencing (I), and moving average (MA) components that repeat at a fixed seasonal frequency s.

Here:

- p, d, q: Regular ARIMA components for non-seasonal patterns.
- P, D, Q: Seasonal counterparts to AR, I, and MA terms.
- s: Length of the seasonal cycle (e.g., 12 for monthly data with yearly seasonality).

SARIMA captures both short-term dynamics and long-term seasonal effects, making it well-suited for time series data exhibiting repeating patterns. It is commonly applied in fields such as retail sales forecasting, energy consumption analysis, climate modeling, and financial market seasonality. Model selection and validation often involve techniques like grid search over hyperparameters, residual diagnostics, and cross-validation on rolling windows to ensure robust performance.

16.5 Operations Science

Operations management is the discipline concerned with designing, overseeing, and improving the production and delivery of goods and services. It plays a pivotal role in maximizing efficiency, minimizing waste, and ensuring the optimal use of resources across various operational contexts. Statistics serve as a critical tool in operations management by supporting data-driven decision-making in areas such as forecasting demand, managing inventory, allocating resources, and maintaining quality standards. This section explores key statistical and mathematical models used in operations management, offering powerful frameworks for solving real-world logistical and organizational challenges.

16.5.1 Queuing Theory

Queuing theory is a branch of operations research that focuses on analyzing systems involving waiting lines or queues. It provides mathematical models to predict queue lengths and waiting times, allowing organizations to design more efficient service processes.

- **Kendall Notation:** A standard notation used to describe queuing systems, such as M/M/1 (where arrivals follow a Poisson process, service times are exponentially distributed, and there is one server).
- **Performance Metrics:** Average queue length, average waiting time, system utilization, and probability of encountering a full system.
- **Multi-Channel Queues:** Extensions like M/M/c for systems with multiple service providers.
- **Use Case:** Evaluating customer wait times at a bank, optimizing call center staffing, improving hospital triage systems, or determining the ideal number of servers at fast-food restaurants.
- **Advantages:** Enables cost-benefit analysis of

adding resources, identifies system bottlenecks, improves customer satisfaction.

- **Limitations:** Requires assumptions about arrival and service time distributions; real systems may not follow idealized models.

16.5.2 Inventory Models

Inventory models are mathematical frameworks used to determine optimal inventory policies that balance ordering and holding costs. They help businesses minimize costs while ensuring product availability.

- **EOQ (Economic Order Quantity):** Calculates the optimal order quantity that minimizes the sum of ordering and holding costs.
  - *Formula:*
    $$EOQ = \sqrt{\frac{2DS}{H}}$$
    where D = annual demand, S = ordering cost per order, and H = annual holding cost per unit.
- **Reorder Point:** The inventory level at which a new order should be placed to avoid stockouts, often based on lead time and demand rate.
- **Safety Stock:** Extra inventory held to protect against variability in demand or lead time.
- **Use Case:** Managing stock levels in a warehouse, scheduling reorders for pharmaceuticals in a

hospital, or optimizing supply chains in retail operations.

- **Advantages:** Reduces costs, improves service levels, supports demand planning.
- **Limitations:** Assumes demand and lead time are known and constant, may not account for all types of uncertainty.

16.5.3 Linear Programming

Linear programming (LP) is a powerful mathematical technique used to optimize a linear objective function subject to a set of linear constraints. It is widely applied in resource allocation, production planning, and transportation.

- **Formulation:** Objective function (maximize or minimize) + constraints (equalities/inequalities) + non-negativity conditions.
- **Simplex Method:** An algorithm that systematically searches for the optimal solution by moving along the edges of the feasible region.
- **Graphical Method:** Visualization tool for solving two-variable LP problems.
- **Duality:** Every LP problem has an associated dual problem; analyzing both can yield insights.
- **Use Case:** Allocating labor hours across production lines, determining optimal product mixes, managing transportation logistics, or

minimizing waste in manufacturing.

- **Advantages:** Provides precise, optimal solutions; supports large-scale decision-making.
- **Limitations:** Assumes linearity and deterministic inputs; may not apply well to non-linear or highly uncertain environments.

16.5.4 Simulation Modeling

Simulation modeling is a technique for analyzing complex systems by imitating their operation over time. It is particularly useful for systems involving uncertainty, variability, and dynamic interactions that are difficult to analyze analytically.

- **Monte Carlo Simulation:** Uses repeated random sampling to model the probability of different outcomes.
- **Discrete-Event Simulation:** Models the system as a sequence of events in time, each of which changes the system's state.
- **Stochastic Simulation:** Incorporates randomness in model parameters to reflect real-world uncertainty.
- **Use Case:** Forecasting supply chain disruptions, assessing risks in project timelines, simulating patient flow in emergency rooms, or evaluating performance of inventory systems under variable demand.

- **Advantages:** Captures real-world complexity; allows for sensitivity analysis; aids in decision-making under uncertainty.
- **Limitations:** Requires computational resources; model accuracy depends on the quality of input data and assumptions; results may be hard to interpret.

Operations management models provide structured approaches to solving resource allocation, scheduling, and process optimization problems. By integrating these models with statistical methods, organizations can transform operational challenges into quantifiable problems, enabling systematic solutions and long-term performance improvements.

16.6 Conclusion

Nonparametric methods offer statistical tools for situations that defy assumptions of normality or linearity. Time series analysis provides powerful insight into data collected over time, enabling forecasts critical for planning. ARIMA and its variants help predict future values with rigor. Operations management applies statistical tools to optimize efficiency and ensure robust decision-making in real-world business systems. Mastering these techniques enables analysts to adapt to diverse challenges and deliver actionable insights.

Chapter 17

Final Chapter

The final chapter brings us to the frontier of analytics: advanced data science and artificial intelligence (AI). These domains extend classical statistical methods into powerful algorithmic systems that learn from data, detect patterns, make predictions, and drive decision-making at scale. Here we explore key model types and tools that define this modern landscape.

17.1 Class Models

Classification models are a core category of supervised learning algorithms designed to assign predefined labels or categories to data points based on their input features. These models learn from labeled training data by identifying statistical patterns, decision boundaries, and relationships that distinguish one class from another. Once trained, classification models predict the category of new, unseen instances with varying levels of confidence. Their applications span numerous domains, including healthcare, finance, cybersecurity, marketing, and natural language processing.

Classification problems may be:

- **Binary:** Two possible outcomes (e.g., spam vs. not spam)
- **Multi-class:** More than two categories (e.g., cancer type classification)
- **Multi-label:** Each observation can belong to multiple classes simultaneously (e.g., tagging an image with several attributes)

The success of classification depends heavily on data quality, appropriate preprocessing, and the selection of a suitable model for the problem.

- **Common Algorithms:**
  - **Logistic Regression:** Estimates the probability of a binary outcome using a logistic function.
  - **Decision Trees:** Uses feature-based splits in a tree structure to form decision rules.
  - **Support Vector Machines (SVM):** Identifies the optimal hyperplane that best separates classes.
  - **k-Nearest Neighbors (k-NN):** Classifies instances based on the majority class among

their k closest neighbors.

  - **Naive Bayes:** Applies Bayes' theorem assuming independence between features.
  - **Random Forest:** An ensemble of decision trees constructed using bagging to improve performance and reduce overfitting.
  - **Gradient Boosted Machines (GBM):** Builds trees sequentially to correct previous errors.
  - **XGBoost and LightGBM:** Scalable, optimized gradient boosting variants for high-dimensional data.

- **Use Cases:**
  - **Email Spam Detection:** Classifies messages as spam or legitimate.
  - **Disease Diagnosis:** Predicts medical conditions based on patient features.
  - **Credit Risk Assessment:** Assesses the likelihood of loan default.
  - **Image Recognition:** Identifies objects, people, or scenes in images.
  - **Sentiment Analysis:** Determines the tone of text as positive, negative, or neutral.
  - **Fraud Detection:** Flags suspicious transactions in finance and insurance.
  - **Customer Churn Prediction:** Estimates

the probability a user will stop using a product or service.

  - **Voice Command Classification:** Recognizes spoken commands using audio signal features.

- **Performance Metrics:**
  - **Accuracy:** Overall rate of correct predictions.
  - **Precision:** Proportion of correctly identified positive instances out of all predicted positives.
  - **Recall (Sensitivity):** Proportion of actual positives that were correctly identified.
  - **F1 Score:** Harmonic mean of precision and recall; helpful in imbalanced datasets.
  - **Confusion Matrix:** A table showing counts of true/false positives and true/false negatives.
  - **ROC Curve and AUC:** Graphical evaluation of true positive vs. false positive rate across thresholds.
  - **Log Loss:** Penalizes incorrect classifications with high confidence.
  - **Cohen's Kappa:** Measures agreement between predicted and true labels, adjusting for chance.

In summary, classification models are essential for problems involving categorization or detection. Their success depends not only on selecting the right algorithm, but also on rigorous data preparation and performance evaluation using appropriate metrics. These models form the foundation for intelligent systems in real-world, high-impact applications.

17.2 Clustering Models

Clustering models are unsupervised learning algorithms that group similar observations together without the use of predefined labels. These models are essential for exploring the hidden structure of data, identifying natural groupings, and discovering underlying patterns in complex datasets. Unlike classification, which predicts known outcomes, clustering reveals unknown relationships, making it a powerful tool for exploratory data analysis.

Common Clustering Algorithms

- **k-Means Clustering:** Partitions data into k clusters by minimizing the variance within each cluster. The algorithm iteratively assigns data points to the nearest centroid and updates centroids until convergence. It is efficient for large datasets but sensitive to outliers and the initial choice of k.
- **Hierarchical Clustering:** Builds a tree-like structure (dendrogram) by either merging or

splitting clusters based on distance metrics. Variants include agglomerative (bottom-up) and divisive (top-down) approaches. Useful for understanding data hierarchies.

- **DBSCAN (Density-Based Spatial Clustering of Applications with Noise):** Groups together points that are closely packed and marks points in low-density regions as outliers. It excels at discovering clusters of arbitrary shape and handling noise.

- **Gaussian Mixture Models (GMM):** Assumes data is generated from a mixture of several Gaussian distributions. It estimates the parameters of these distributions using the Expectation-Maximization (EM) algorithm and assigns probabilistic cluster memberships.

Key Applications

- **Market Segmentation:** Identifying distinct customer groups based on purchasing behavior or demographics.

- **Anomaly Detection:** Detecting fraudulent activities, network intrusions, or defective products by isolating outliers.

- **Image Compression:** Reducing file size by grouping similar pixels or color values.

- **Document Categorization:** Automatically grouping similar text documents for topic modeling or summarization.

- **Recommender Systems:** Clustering users or items to improve collaborative filtering and personalization.

Evaluation Techniques

Evaluating clustering results is challenging due to the absence of ground truth labels. Nonetheless, several internal and external metrics are used:

- **Silhouette Score:** Measures how similar an object is to its own cluster compared to others. Values range from -1 to 1, with higher values indicating better clustering.
- **Dunn Index:** Assesses cluster compactness and separation. Higher values suggest well-separated clusters.
- **Within-Cluster Sum of Squares (WCSS):** Used primarily in k-Means to assess compactness. The Elbow Method plots WCSS to find the optimal k.
- **Adjusted Rand Index (ARI):** Compares clustering results to ground truth labels when available.
- **Cluster Stability:** Tests the robustness of clusters under resampling or perturbations.

Challenges in Clustering

- **Choosing the Right Algorithm:** Different algorithms perform better on different types of data (e.g., density-based vs. centroid-based).
- **Determining Number of Clusters:** Many algorithms require a predefined number of clusters, which may not be obvious.
- **Scalability:** Algorithms like hierarchical clustering can be computationally intensive for large datasets.
- **Interpretability:** The meaning of clusters must be inferred and validated by domain knowledge.
- **Sensitivity to Scaling and Noise:** Clustering results can be significantly influenced by data scaling and outliers.

Best Practices

- **Preprocess Your Data:** Clean and sanitize the data. Normalize features, remove outliers, and consider dimensionality reduction (e.g., PCA).
- **Try Multiple Algorithms:** Compare results from different clustering techniques to understand the data structure better.
- **Visualize Clusters:** Use tools like t-SNE or PCA for dimensionality reduction and cluster visualization.

- **Combine with Supervised Learning:** Cluster assignments can be used as features or for semi-supervised learning.
- **Use Domain Knowledge:** Validate cluster quality and interpretability based on the application context.

In conclusion, clustering models offer a powerful suite of tools for uncovering hidden structures in data. From customer insights to anomaly detection, these models provide foundational support in exploratory analytics, guiding deeper inquiry and strategic decision-making.

17.3 Time Series

Time series forecasting is a specialized branch of statistical and machine learning analysis focused on predicting future values based on previously observed temporal data. These models are foundational in areas such as economics, finance, supply chain, energy demand, weather prediction, and healthcare operations. By leveraging the structure inherent in time-stamped data, forecasters can model trends, seasonality, cycles, and random noise to make informed predictions about what comes next.

Key Concepts in Time Series Analysis

- **Trend:** Long-term increase or decrease in the data.
- **Seasonality:** Repeating patterns or cycles over

fixed intervals (e.g., daily, monthly).

- **Cyclicality:** Irregular rises and falls not tied to a specific period.
- **Stationarity:** A stationary time series has constant mean, variance, and autocorrelation over time. Many forecasting methods assume stationarity.
- **Noise:** Random variation or error that cannot be explained by the model.

Forecasting Models and Techniques

- **ARIMA (AutoRegressive Integrated Moving Average):** One of the most widely used classical forecasting models. It consists of:
  - $AR(p)$: Autoregressive part—uses past values to predict the current value.
  - $I(d)$: Integration—differencing the time series d times to make it stationary.
  - $MA(q)$: Moving average—uses past forecast errors to improve current predictions.

  Steps:

  1. Plot and visualize the series.
  2. Use Augmented Dickey-Fuller (ADF) test to assess stationarity.
  3. Apply differencing if needed.

4. Examine ACF and PACF plots to choose p and q.

5. Fit ARIMA using methods such as maximum likelihood estimation.

6. Diagnose residuals to verify randomness (e.g., using Ljung-Box test).

7. Forecast and compute prediction intervals.

- **Exponential Smoothing Models:** Focuses on smoothing past observations, applying decreasing weights to older values.

  - **Simple Exponential Smoothing (SES):** Appropriate for data with no clear trend or seasonality.
  - **Holt's Linear Trend Method:** Extends SES by adding trend.
  - **Holt-Winters Method:** Adds seasonal adjustments for both additive and multiplicative seasonality.

  These methods are fast and interpretable and provide good benchmarks.

- **Facebook Prophet:** Designed for business users with minimal statistical background. Prophet decomposes the time series into trend, seasonality, and holiday effects using an additive model. It automatically detects changepoints and is robust to missing data and outliers.

  - Suitable for daily/weekly business time series

with strong seasonal effects.

  - Provides intuitive tunability via changepoint prior scale and seasonal modes.
  - Outputs include forecast plots with uncertainty intervals.

- **Temporal Fusion Transformer (TFT):** A cutting-edge deep learning model designed for high-performance multi-horizon forecasting.
  - Combines LSTM and attention mechanisms.
  - Handles static covariates, time-varying inputs, and known future inputs.
  - Provides variable importance explanations.
  - Used in industrial and research contexts for highly complex datasets.

Model Selection and Evaluation

- **Train-Test Split:** Maintain chronological integrity; don't shuffle.
- **Cross-Validation:** Use rolling forecast origin or time-series split.
- **Evaluation Metrics:**
  - Mean Absolute Error (MAE)
  - Root Mean Square Error (RMSE)
  - Mean Absolute Percentage Error (MAPE)

    - Symmetric MAPE (sMAPE)
- **Residual Analysis:** Ensure residuals resemble white noise.

Best Practices for Forecasting

- Always visualize your time series to identify structure and outliers.
- Start with simple models (e.g., naive forecast, moving average) as baselines.
- Consider data transformations (log, Box-Cox) to stabilize variance.
- Validate forecast accuracy using out-of-sample tests.
- Combine models for ensemble forecasting if no single model excels.

Applications of Time Series Forecasting

- **Retail:** Demand forecasting for inventory and supply chain.
- **Finance:** Stock price prediction, volatility estimation.
- **Energy:** Load forecasting for energy production and distribution.
- **Healthcare:** Predicting patient inflow, epidemic spread.

- **Weather:** Forecasting temperature, precipitation, storm probability.

Time series forecasting enables organizations to move from reactive to proactive decision-making. By modeling and predicting future states, analysts can guide strategy, resource allocation, and risk mitigation.

17.4 Deep Learning

Deep learning is a subfield of machine learning that leverages multi-layered artificial neural networks to model and understand intricate structures in vast and high-dimensional datasets. Inspired loosely by the human brain, deep learning models are capable of automatically learning representations from raw data, reducing the need for manual feature engineering. With the explosion of data and advancements in computing hardware, particularly GPUs and TPUs, deep learning has emerged as a powerful tool for a wide array of tasks across industries.

What Makes Deep Learning Unique

Deep learning differs from traditional machine learning in its ability to handle raw, unstructured data (e.g., images, text, audio) and discover abstract representations through a hierarchy of layers. Each layer in a deep network learns features at a different level of abstraction, from simple edges in early layers to complex object representations in deeper ones. This layered architecture allows deep learning models to excel in tasks previously thought to be solvable only through human intuition.

Core Neural Network Architectures

- **Feedforward Neural Networks (FNN):** The simplest architecture where information moves in one direction from input to output without cycles.
- **Convolutional Neural Networks (CNN):** Primarily used for image and video processing. They apply convolutional layers to capture spatial hierarchies.
- **Recurrent Neural Networks (RNN):** Designed for sequential data like time series and natural language. Variants like LSTM and GRU improve long-term memory.
- **Transformers:** Revolutionized NLP and time-series forecasting by using attention mechanisms instead of recurrence. BERT and GPT are based on this architecture.
- **Autoencoders:** Used for unsupervised learning and dimensionality reduction. They reconstruct inputs from compressed representations.
- **Generative Adversarial Networks (GANs):** Consist of a generator and discriminator competing to create realistic synthetic data, widely used in image generation and data augmentation.

Popular Deep Learning Frameworks

- **TensorFlow:** An end-to-end open-source platform developed by Google, suitable for building and deploying large-scale models.
- **PyTorch:** Developed by Facebook, it offers dynamic computation graphs and is favored for research and experimentation.
- **Keras:** A high-level API running on top of TensorFlow, designed for rapid prototyping and ease of use.

Applications Across Domains

- **Computer Vision:** Image classification, object detection, facial recognition, autonomous vehicles.
- **Natural Language Processing (NLP):** Language translation, sentiment analysis, question answering, text generation.
- **Speech Processing:** Voice recognition, speech-to-text, text-to-speech, speaker identification.
- **Healthcare:** Medical image analysis, drug discovery, disease diagnosis.
- **Finance:** Fraud detection, algorithmic trading, credit scoring.
- **Gaming and Robotics:** Reinforcement learning for autonomous agents and real-time decision

making.

- **Art and Creativity:** Generative art, music composition, deepfake videos.

Challenges in Deep Learning

Despite its transformative capabilities, deep learning presents several challenges:

- **Data Requirements:** Training effective deep models requires vast amounts of labeled data.
- **Computational Demands:** Deep networks are resource-intensive, requiring powerful GPUs and extensive training time.
- **Model Interpretability:** The complexity of deep models often makes them opaque, leading to challenges in understanding how decisions are made (black-box problem).
- **Overfitting:** With high capacity to learn, deep networks can memorize training data without generalizing well to new data. Regularization methods like dropout, weight decay, and data augmentation are commonly used.
- **Ethical and Bias Concerns:** If trained on biased data, models can perpetuate societal inequalities. Transparency, fairness, and accountability must be incorporated.

Best Practices for Deep Learning

- **Data Preprocessing:** Normalize, augment, and clean the data before feeding into models.
- **Model Architecture Tuning:** Select appropriate depth, width, and activation functions for your task.
- **Use Pretrained Models:** Leverage transfer learning from models like ResNet, BERT, or GPT to reduce training time and data requirements.
- **Evaluation and Validation:** Use validation sets, cross-validation, and performance metrics like AUC, F1-score, and confusion matrices.
- **Monitoring and Deployment:** Ensure models are monitored post-deployment for drift, fairness, and accuracy.

In conclusion, deep learning represents the cutting edge of artificial intelligence. Its ability to uncover nuanced, abstract patterns from raw data makes it indispensable in modern data science. However, its power must be matched with ethical considerations, careful design, and proper evaluation to ensure responsible and impactful outcomes.

17.5 Machine Learning

Machine learning is a foundational pillar of modern data science, encompassing a broad spectrum of algorithms and methodologies that enable computers to learn from data and make intelligent decisions with minimal human intervention. It lies at the intersection of computer science, statistics, and optimization, and powers a vast array of real-world applications, from predicting stock prices to detecting credit card fraud. Machine learning algorithms can be broadly categorized into supervised, unsupervised, semi-supervised, and reinforcement learning, each suited to different types of tasks and data scenarios.

Categories of Machine Learning

- **Supervised Learning:** Algorithms learn from labeled data. Examples include linear regression, logistic regression, support vector machines (SVM), decision trees, and ensemble models like random forest and gradient boosted machines.
- **Unsupervised Learning:** Algorithms uncover hidden patterns in data without labels. Techniques include clustering (e.g., k-Means, DBSCAN), dimensionality reduction (e.g., PCA), and anomaly detection.
- **Semi-supervised Learning:** Combines a small amount of labeled data with a large amount of unlabeled data, improving learning efficiency.
- **Reinforcement Learning:** An agent learns to make decisions by receiving rewards or penalties from interactions with its environment. Popular in robotics, gaming, and real-time decision-making.

Key Algorithms and Techniques

- **Linear Regression:** Predicts continuous values by modeling linear relationships between input features and output.
- **Logistic Regression:** Classifies binary outcomes using a logistic function.
- **Decision Trees and Random Forests:** Tree-based models used for classification and regression. Random Forests mitigate overfitting through bagging.
- **Support Vector Machines (SVM):** Finds the hyperplane that best separates classes in high-dimensional spaces.
- **k-Nearest Neighbors (k-NN):** A non-parametric method that assigns a class based on the majority class among its nearest neighbors.
- **Gradient Boosted Machines (GBM) and XGBoost:** Ensemble methods that combine weak learners to build strong predictive models.
- **Naive Bayes:** Probabilistic classifier based on Bayes' Theorem with strong (naive) independence assumptions.

Automated Machine Learning (AutoML)

AutoML automates the process of building machine learning pipelines, including data preprocessing, feature engineering, model selection, and hyperparameter tuning. It democratizes AI by making model building accessible to non-experts. Key platforms include:

- **H2O AutoML:** An open-source platform offering automated training and ensembling of models.
- **Google Cloud AutoML:** Provides user-friendly interfaces for training high-quality models without extensive coding.
- **Azure AutoML:** Offers model explainability and integration with Azure ML pipelines.
- **TPOT:** A genetic programming-based approach for optimizing machine learning pipelines in Python.

Popular Tools and Libraries

- **Scikit-learn:** A robust Python library providing tools for classification, regression, clustering, and preprocessing.
- **H2O.ai:** Offers scalable machine learning and deep learning capabilities.
- **RapidMiner:** A visual workflow platform for data science.
- **Microsoft Azure ML Studio:** A cloud-based platform for drag-and-drop machine learning model development.
- **MLlib (Spark):** Apache Spark's scalable machine learning library for large datasets.

Model Evaluation and Validation

- **Cross-validation:** Evaluates model performance by partitioning data into training and validation folds.
- **Confusion Matrix:** Tabulates true vs. predicted classifications.
- **Metrics:** Includes accuracy, precision, recall, F1-score, ROC-AUC, and RMSE for regression tasks.
- **Grid Search and Random Search:** Techniques for tuning hyperparameters to optimize model performance.
- **Bias-Variance Tradeoff:** Balance between underfitting and overfitting.

Challenges in Machine Learning

- **Data Quality and Quantity:** Insufficient or biased data leads to poor generalization.
- **Overfitting:** Model performs well on training data but fails to generalize.
- **Model Interpretability:** Complex models like ensembles and neural networks are often black boxes.
- **Fairness and Ethics:** Ensuring models do not perpetuate or amplify bias present in data.

- **Scalability:** Efficiently training and deploying models in distributed environments.

Best Practices in Machine Learning

- **Data Preprocessing:** Impute missing values, encode categorical variables, and scale numeric features.
- **Feature Engineering:** Create meaningful features that capture underlying patterns.
- **Regularization:** Techniques like Lasso and Ridge to prevent overfitting.
- **Interpretability Tools:** SHAP, LIME, and model-agnostic techniques for explaining predictions.
- **Continuous Monitoring:** Evaluate deployed models for concept drift and performance degradation.

In summary, machine learning is a versatile and evolving field that powers intelligent systems across virtually every sector. Its continual progress depends on the careful blending of domain knowledge, statistical rigor, computational power, and ethical responsibility.

17.6 Big Data

Big data technologies provide the foundation for storing, processing, and analyzing vast volumes of data that are too large, fast, or complex for traditional systems. These technologies are crucial for modern data science, enabling real-time insights, scalability, and integration across distributed systems and cloud platforms. The rise of big data has transformed industries by allowing businesses and researchers to harness data-driven decision-making at scale.

Defining Characteristics of Big Data

Big data is typically defined by the "3Vs":

- **Volume:** Refers to the massive scale of data generated from sources such as IoT devices, social media, sensors, transactions, and web logs.
- **Velocity:** The speed at which new data is generated and must be processed, often requiring real-time or near-real-time systems.
- **Variety:** The different formats and structures of data, including structured (databases), semi-structured (XML, JSON), and unstructured (text, images, videos).
- **Additional Vs:** Some definitions add Veracity (uncertainty and data quality) and Value (business usefulness).

Key Technologies and Tools

- **Hadoop:** An open-source framework using the MapReduce programming model for processing large datasets across clusters of computers.

- **Spark:** A fast and general-purpose cluster computing system that improves on Hadoop by enabling in-memory data processing.
- **Hive:** A data warehouse infrastructure built on top of Hadoop that provides SQL-like queries.
- **Presto:** A high-performance distributed SQL engine for querying big data across sources.
- **Flink:** A stream-processing framework that supports batch and real-time data analytics.
- **Kafka:** A distributed event streaming platform used for building real-time data pipelines.

Big Data Architecture

Big data architecture typically includes:

- **Data Ingestion:** Collecting data from multiple sources (e.g., web apps, sensors, logs).
- **Data Storage:** Using HDFS, cloud storage (S3, Azure Blob), or NoSQL databases (e.g., Cassandra, MongoDB).
- **Data Processing:** Batch (e.g., MapReduce, Spark) and stream processing (e.g., Flink, Storm).
- **Data Analysis:** Tools like Hive, Presto, or Python/R-based libraries.
- **Data Visualization:** Integrating with tools like Tableau, Power BI, or D3.js.

Use Cases and Applications

Big data technologies are used across industries for a wide range of applications:

- **Healthcare:** Analyzing patient records and genomics for precision medicine.
- **Finance:** Real-time fraud detection, algorithmic trading, and risk modeling.
- **Retail:** Customer sentiment analysis, recommendation systems, and dynamic pricing.
- **Manufacturing:** Predictive maintenance and supply chain optimization.
- **Social Media and Marketing:** Trend analysis and targeted ad campaigns.

Challenges in Big Data

- **Data Quality:** Ensuring accuracy, consistency, and completeness of massive datasets.
- **Scalability:** Managing data growth across systems without sacrificing performance.
- **Security and Privacy:** Protecting sensitive data in compliance with regulations like GDPR and HIPAA.
- **Integration:** Combining data from disparate sources with varying formats.

- **Skill Gap:** Need for professionals proficient in big data tools, architecture, and analysis.

Emerging Trends

- **Cloud-native Big Data:** Use of cloud platforms (AWS, Azure, GCP) for scalable data lakes and warehouses.
- **Edge Computing:** Processing data closer to the source to reduce latency.
- **AI Integration:** Combining big data with AI/ML for predictive analytics and automation.
- **Data Fabric and Mesh:** New paradigms for managing decentralized data across environments.

In summary, big data technologies provide the backbone for handling today's data explosion. With the right infrastructure and strategies, organizations can turn large-scale, complex datasets into actionable insights that drive innovation and competitive advantage.

17.7 NLP

Natural Language Processing (NLP) enables computers to understand, interpret, and generate human language.

- **Techniques:** Tokenization, stemming, lemmatization, part-of-speech tagging, named entity recognition.
- **Models:** BERT, GPT, LSTM, Transformer

architectures.

- **Applications:** Chatbots, sentiment analysis, text summarization, language translation.

17.8 Applications

Advanced data science applications span virtually every sector and play a pivotal role in transforming decision-making, process efficiency, customer relations, and risk management. These models harness structured and unstructured data, combining machine learning, statistical techniques, and domain-specific knowledge to uncover insights and enable proactive actions. The utility of these models is enhanced by integration into modern technological ecosystems, including cloud computing, edge devices, and real-time analytics platforms.

Strategic Decision-Making

- **Forecasting and Planning:** Use predictive models to estimate future trends in sales, demand, inventory, or staffing, allowing better allocation of resources and long-term strategic planning.
- **Scenario Simulation:** Conduct what-if analyses using simulation or optimization models to evaluate the outcomes of different decisions or environmental conditions.
- **Real-Time Dashboards:** Integrate model outputs with business intelligence tools for live

monitoring and agile responses.

- **Use Cases:** Retailers forecasting seasonal sales, hospitals planning emergency staffing, airlines scheduling flights based on dynamic demand.

Operational Optimization

- **Resource Allocation:** Optimize how assets such as trucks, technicians, or call center staff are assigned using linear programming and heuristic algorithms.
- **Process Improvement:** Analyze workflows to reduce bottlenecks, idle time, or quality issues through Six Sigma, time series, and control charts.
- **Scheduling Algorithms:** Use data-driven scheduling tools to balance supply chains, manufacturing runs, or healthcare appointments.
- **Use Cases:** Factory assembly line sequencing, telecom tower maintenance routing, hospital bed and staff utilization.

Customer Experience Enhancement

- **Personalization Engines:** Leverage collaborative filtering and deep learning models to tailor recommendations, search results, or promotions.
- **Chatbots and NLP Systems:** Provide human-like automated support, reducing customer

service costs while improving accessibility.

- **Sentiment Analysis:** Use natural language processing to understand public sentiment toward brands, products, or political figures.
- **Use Cases:** Netflix movie recommendations, Amazon product promotions, Spotify playlist generation, bank chatbot services.

Risk Prediction and Prevention

- **Fraud Detection:** Use anomaly detection and supervised learning on transaction data to catch fraudulent activities in real time.
- **Credit Risk Assessment:** Apply classification models to determine loan eligibility and expected default probability.
- **Predictive Maintenance:** Anticipate machine failure using sensor data and time series analytics to reduce downtime.
- **Use Cases:** Insurance underwriting, cybersecurity alert systems, automotive telematics.

Scientific and Social Innovation

- **Precision Medicine:** Analyze genetic and patient health data to tailor treatments specific to individual biological profiles.

- **Climate Modeling:** Forecast climate change and disaster patterns using large-scale simulation and geospatial data.
- **Urban Analytics:** Leverage real-time city data to optimize traffic, utilities, and public health interventions.
- **Use Cases:** COVID-19 spread prediction, earthquake detection systems, pollution monitoring.

Integrated Applications Across Ecosystems

- **Cross-Platform Integration:** Advanced models are embedded into mobile apps, cloud APIs, IoT devices, and enterprise ERPs.
- **Edge Analytics:** Deploy models on edge devices (e.g., drones, wearables) for low-latency, autonomous decision-making.
- **Cloud Pipelines:** Automate training, validation, and deployment using MLOps workflows across platforms like AWS, GCP, and Azure.

In essence, advanced data science applications are indispensable tools for navigating complexity in today's world. They deliver actionable insights, uncover hidden opportunities, and create measurable impact across nearly every domain—from corporate boardrooms to scientific laboratories, and from personalized digital experiences to global humanitarian interventions.

17.9 Key Platforms

The modern landscape of artificial intelligence and machine learning is supported by a diverse set of platforms and tools that empower data scientists, engineers, and businesses to develop, train, validate, and deploy advanced models at scale. These platforms span commercial cloud ecosystems, open-source toolkits, and integrated development environments (IDEs) designed for collaborative analytics, scalable computation, and robust deployment.

Cloud AI Platforms

- **Amazon SageMaker:** A fully managed service offered by AWS that enables developers and data scientists to build, train, and deploy machine learning models quickly. SageMaker provides built-in algorithms, AutoML functionality, and integrated Jupyter notebooks, as well as support for deep learning frameworks such as TensorFlow, PyTorch, and MXNet. The platform also includes SageMaker Studio for visual development, and tools for MLOps, including model monitoring and CI/CD pipelines.
- **Google Cloud Vertex AI:** A unified machine learning platform that integrates AutoML and custom model training with Google's suite of data and AI services. It supports scalable workflows for supervised, unsupervised, and deep learning models. Vertex AI offers pre-built models, model

explainability tools, pipeline orchestration, and integration with BigQuery ML and TensorFlow Extended (TFX).

- **Azure Machine Learning:** Microsoft's enterprise-grade platform for developing, deploying, and managing ML models using both no-code and code-first approaches. It includes automated machine learning, data labeling, responsible AI toolkits for fairness and interpretability, and seamless integration with Azure Databricks and other cloud services. Users can scale experiments across GPUs and manage entire lifecycles using Azure ML Pipelines.
- **Databricks Data Intelligence Platform:** Built on Apache Spark, Databricks unifies data engineering, data science, and machine learning. It supports collaborative notebooks, distributed computing, and MLflow for experiment tracking. The platform is ideal for large-scale data processing, model training, and AI-driven business intelligence.

AutoML Platforms

- **H2O Driverless AI:** An AutoML platform that automates feature engineering, model training, hyperparameter tuning, and interpretability. It supports time series forecasting, classification, and regression tasks with model explanation tools such as SHAP and LIME.

- **Google AutoML:** Designed for users with limited ML expertise, this tool allows rapid development of image, video, and text classification models through transfer learning and automated pipeline construction.
- **AWS Autopilot:** Integrated into SageMaker, it simplifies ML by automatically preprocessing data, training models, and selecting the best-performing model based on the desired metrics.

Open-Source Frameworks and Libraries

- **TensorFlow and Keras:** Widely used for building deep learning models, these libraries offer modular APIs for designing, training, and deploying complex neural networks.
- **PyTorch:** Popular in research and production settings, it provides dynamic computational graphs, seamless GPU acceleration, and a growing ecosystem of pretrained models and libraries such as TorchVision and HuggingFace Transformers.
- **Scikit-learn:** A general-purpose ML library for Python that includes efficient implementations of classification, regression, clustering, dimensionality reduction, and model evaluation.
- **XGBoost and LightGBM:** Specialized gradient boosting libraries that offer state-of-the-art performance for structured/tabular data.

Big Data and Distributed Processing Tools

- **Apache Hadoop:** A framework for distributed storage (HDFS) and batch processing (MapReduce), useful for handling large volumes of unstructured data.
- **Apache Spark:** A fast, general-purpose engine for big data processing with built-in modules for SQL, streaming, MLlib (machine learning), and GraphX (graph processing).
- **Apache Flink and Apache Beam:** Stream processing tools ideal for real-time analytics and large-scale event-driven applications.
- **Presto and Hive:** SQL-based engines for querying big data stored in distributed file systems.

Integrated Development Environments (IDEs) and Collaboration Platforms

- **JupyterLab:** A flexible and interactive environment for writing code, running experiments, visualizing data, and sharing notebooks across teams.
- **VS Code with AI Extensions:** Supports Python, R, Julia, and other languages with plugins for TensorFlow, PyTorch, and remote kernel execution.

- **Google Colab:** Free, browser-based notebooks with GPU access and direct integration with Google Drive, ideal for rapid prototyping and education.
- **Kaggle Kernels:** An online environment for collaborative data science projects, often paired with competitions and public datasets.

MLOps and Deployment Tools

- **MLflow:** An open-source platform for managing the end-to-end machine learning lifecycle, including experiment tracking, model versioning, and deployment.
- **Kubeflow:** A Kubernetes-native platform for running scalable and portable ML workflows.
- **TensorFlow Serving and TorchServe:** Model serving frameworks that allow real-time inference via RESTful APIs.
- **Docker and Kubernetes:** Widely used for containerization and orchestration of scalable ML applications.

These platforms and tools together form the technological backbone of modern AI and data science workflows. They allow organizations to scale experimentation, accelerate deployment, automate repetitive tasks, and build robust, production-ready intelligent systems.

17.10 AI Meets Data Science

Artificial Intelligence (AI) and Data Science, though often delineated as separate domains, converge dynamically to form the foundational framework of intelligent systems in the modern digital era. At their core, data science provides the rigorous statistical and computational methodologies to collect, clean, analyze, and visualize data, while AI, particularly through its machine learning and deep learning subfields, develops autonomous models that learn from data to mimic or augment human cognition and decision-making. When intertwined, these fields not only complement but also amplify each other's capabilities, transforming vast data lakes into functional intelligence.

AI applications, whether in voice assistants, fraud detection, healthcare diagnostics, or autonomous vehicles, depend heavily on the data infrastructure, analytics, and domain expertise cultivated by data science. Conversely, the evolution of data science is increasingly marked by the integration of AI-powered automation tools that accelerate data processing, feature engineering, model selection, and hyperparameter tuning. The synergy yields systems capable of real-time adaptation, contextual learning, and large-scale prediction that were once impossible with traditional statistical models alone.

This convergence is not only technical but also strategic. AI and data science teams collaborate across industries to design solutions that integrate data pipelines with AI models and production environments.

With the advent of MLOps (Machine Learning Operations), the lifecycle of a model—from training to deployment to monitoring—is becoming tightly woven with both disciplines, fostering more reliable, scalable, and explainable systems.

Examples of their integration include AI-powered BI dashboards that predict trends in real-time, natural language interfaces trained on domain-specific datasets, and adaptive systems that personalize digital experiences based on continuous data ingestion. Innovations such as AutoML exemplify this union, automating repetitive tasks like algorithm selection and parameter tuning, thus freeing data scientists to focus on strategy, interpretation, and business impact.

Key Synergies Between AI and Data Science

- **AI-enhanced Data Science:** Algorithms accelerate data wrangling, visualization, forecasting, and causal inference. AI tools assist in identifying relationships within data that might be overlooked by manual analysis.
- **Data Science-driven AI:** Statistical rigor, domain knowledge, and data literacy shape robust AI models. Preprocessing, feature selection, and interpretability metrics rooted in data science practices increase model reliability and fairness.
- **Unified Ecosystems and Tools:** Shared programming languages (Python, R), cloud

environments (AWS, GCP, Azure), and interactive notebooks (Jupyter, Colab) support collaboration and end-to-end pipeline development.

- **MLOps and CI/CD Pipelines:** Workflow orchestration, containerization, continuous integration, and model monitoring practices ensure models are reproducible, scalable, and well-governed.

- **AutoML and Augmented Analytics:** Automated tools empower users with minimal coding experience to build predictive models, democratizing access to AI and boosting organizational efficiency.

- **AI for Real-Time Analytics:** Coupling streaming data tools with real-time inference engines enables proactive decision-making across domains such as cybersecurity, e-commerce, and IoT.

- **Shared Goals:** Both disciplines aim to extract actionable knowledge, reduce uncertainty, and drive innovation through intelligent use of information.

As the boundaries between AI and data science blur, professionals from both fields are encouraged to develop interdisciplinary fluency. Data scientists benefit from understanding neural architectures and reinforcement learning, while AI engineers gain insights from experimental design and statistical inference. This cross-pollination fuels the next generation of intelligent systems that are not only technically powerful but also ethically grounded and socially beneficial.

In summary, AI and data science are no longer parallel tracks—they are interdependent collaborators that together shape the future of decision-making, automation, and discovery. Their integration defines the core architecture of intelligent digital transformation across sectors.

Conclusion

This chapter closes the book by highlighting the power of advanced data science and artificial intelligence. While the foundational methods from earlier chapters remain essential, today's data professionals must also embrace complex models, high-dimensional data, and scalable technologies. By combining traditional statistical rigor with cutting-edge tools, we build the future of intelligent, data-driven decision-making.

From linear models to neural networks, from t-tests to transformers, the statistical journey continues into deeper waters. Let your insights evolve as your data does.

Appendix A

Answer Key

Answer Key to Chapter 5 Exercises

1. A population is the entire group; a sample is a subset. Example: All registered voters in a country (population), 100 randomly selected voters (sample).
2. (a) Parameter

   (b) Statistic
3. Mean = 5.7, Median = 5, Mode = 4 and 7 (bimodal)
4. The value 95 is an outlier. It inflates the mean significantly but does not affect the median much. The mode is unaffected.
5. Range = 10, Variance = 13.0, Standard Deviation 3.61

6. $\text{CV} = \frac{5}{30} \times 100\% = 16.67\%$

7. Midpoints: 14.5, 24.5, 34.5, 44.5; Mean $\frac{(14.5)(2)+(24.5)(6)+(34.5)(8)+(44.5)(4)}{20} = 32$

8. Positively skewed: the mean is pulled to the right due to the long tail.

9. 5-number summary: Min = 2, Q1 = 5, Median = 10, Q3 = 17, Max = 21. Boxplot is drawn accordingly.

10. - Sample variance: $s^2 = \frac{(6-9.5)^2+(8-9.5)^2+(10-9.5)^2+(14-9.5)^2}{3} = 11$
    - Population variance: $\sigma^2 = \frac{(6-9.5)^2+(8-9.5)^2+(10-9.5)^2+(14-9.5)^2}{4} = 8.25$

Answer Key to Chapter 6 Exercises

1. **Classical Probability:** Even numbers on a die are 2, 4, and 6. There are 3 favorable outcomes out of 6 total outcomes.

$$P(\text{even}) = \frac{3}{6} = 0.5$$

2. **Empirical Probability:**

$$P(\text{not defective}) = \frac{160}{200} = 0.8$$

3. **Subjective Probability:** This is an example of *subjective probability*, as it is based on personal judgment or opinion rather than direct computation or data.

4. **Complement Rule:**

$$P(\text{not rain}) = 1 - 0.35 = 0.65$$

5. **Addition Rule (Mutually Exclusive):**

$$\begin{aligned} P(\text{King or Queen}) &= P(\text{King}) + P(\text{Queen}) \\ &= \frac{4}{52} + \frac{4}{52} \\ &= \frac{8}{52} = \frac{2}{13} \end{aligned}$$

6. **Addition Rule (Not Mutually Exclusive):**

$$P(\text{left or glasses}) = 0.1 + 0.25 - 0.05 = 0.3$$

7. **Multiplication Rule (Independent Events):**

$$P(\text{heads and } 3) = P(\text{heads}) \cdot P(3) = \frac{1}{2} \cdot \frac{1}{6} = \frac{1}{12}$$

8. **Conditional Probability:**

$$P(\text{vanilla} \mid \text{chocolate}) = 0.4$$

9. **Tree Diagram & Conditional Probability:**

$$P(\text{red then red}) = \frac{3}{5} \cdot \frac{2}{4} = \frac{6}{20} = 0.3$$

10. **Bayes' Theorem:** Let $D =$ has disease, $T =$ tests positive.

$$\begin{aligned} P(D \mid T) &= \frac{P(T \mid D) \cdot P(D)}{P(T \mid D) \cdot P(D) + P(T \mid \text{no D}) \cdot P(\text{no D})} \\ &= \frac{0.95 \cdot 0.02}{0.95 \cdot 0.02 + 0.05 \cdot 0.98} \\ &= \frac{0.019}{0.019 + 0.049} \\ &= \frac{0.019}{0.068} \approx 0.279 \end{aligned}$$

Answer Key to Chapter 7 Exercises

1. $\mu = 2.7$, $\sigma^2 = 0.81$, $\sigma = 0.9$.
2. $P(X = 1)+P(X = 2)+P(X = 3)+P(X = 4) = 1$; **Yes**, it is a valid PMF.
3. $\mu = 2.7$, $\sigma^2 = 1.09$, $\sigma \approx 1.044$.
4. $P(X \geq 2) = 0.6$.
5. Binomial Distribution: $P(X = 3) = \binom{5}{3}(0.4)^3(0.6)^2 = 0.2304$.
6. $\mu = 2$, $\sigma^2 = 1.2$, $\sigma \approx 1.095$.
7. $P(X = 0) = 0.1353$ using the Poisson formula.
8. Hypergeometric Distribution:

$$P(X = 2) = \frac{\binom{5}{2}\binom{7}{3}}{\binom{12}{5}} = 0.3968$$

9. $\mu = 3$, $\sigma^2 = 3$, $\sigma = \sqrt{3} \approx 1.732$ (Poisson).

10. $P(X \geq 2) = 1 - P(X = 0) - P(X = 1) = 1 - e^{-3} - 3e^{-3} = 1 - 4e^{-3} \approx 0.8013$.

Answer Key to Chapter 8 Exercises

1. $P(X < 8) = \frac{8-5}{10-5} = \frac{3}{5} = 0.6$

2. $P(X > 2) = \int_2^\infty \lambda e^{-\lambda x} dx = e^{-2} \approx 0.1353$

3. $Z = \frac{85-80}{10} = 0.5 \Rightarrow P(X < 85) = 0.6915$

4. $\mu = \frac{2+6}{2} = 4$, $\sigma^2 = \frac{(6-2)^2}{12} = \frac{16}{12} = \frac{4}{3}$

5. $\mu = \frac{1}{2} = 2$, $\sigma^2 = \frac{1}{\lambda^2} = \frac{1}{4}$

6. $Z = \frac{60-55}{5} = 1 \Rightarrow P(X > 60) = 1 - 0.8413 = 0.1587$

7. $Z = \frac{90-70}{15} = 1.33 \Rightarrow P(X > 90) = 1 - 0.9082 = 0.0918$

8. $\mu = \frac{0+10}{2} = 5$, $\sigma^2 = \frac{(10-0)^2}{12} = \frac{100}{12} \approx 8.33$

9. $P(X < 2) = 1 - e^{-2} \approx 1 - 0.1353 = 0.8647$

10. $Z_1 = \frac{65-60}{8} = 0.625$, $Z_2 = \frac{75-60}{8} = 1.875$
    $P(65 < X < 75) = P(Z < 1.875) - P(Z < 0.625) = 0.9699 - 0.7340 = 0.2359$

Answer Key to Chapter 9 Exercises

1. **Standard Error of Sample Mean:**

$$SE_{\bar{x}} = \frac{\sigma}{\sqrt{n}} = \frac{15}{\sqrt{25}} = \frac{15}{5} = 3$$

2. **Using the z-distribution:**

$$SE = \frac{1.2}{\sqrt{49}} = \frac{1.2}{7} \approx 0.1714$$

$$z = \frac{7 - 6.8}{0.1714} \approx 1.167$$

$$P(\bar{x} > 7) \approx 1 - 0.878 = 0.122$$

3. **Central Limit Theorem:**
Since $n = 40 \geq 30$, the CLT applies.
The sampling distribution of $\bar{x}$ is approximately normal.

4. **Standard Error of Proportion:**

Compute:

$$\hat{p} = 0.48, \quad 1 - \hat{p} = 0.52$$

5. **Standard Error of Proportion:**

We are given $\hat{p} = 0.48$, so:

$$SE_{\hat{p}} = \sqrt{\frac{0.48 \times 0.52}{150}}$$

$$= \sqrt{\frac{0.2496}{150}}$$

$$= \sqrt{0.001664}$$

$$\approx 0.0408$$

6. **Standard Error of Difference Between Means:**

$$SE = \sqrt{\frac{25^2}{60} + \frac{30^2}{55}} = \sqrt{\frac{625}{60} + \frac{900}{55}}$$

$$= \sqrt{10.42 + 16.36} = \sqrt{26.78} \approx 5.17$$

7. ***t*-Statistic:**

$$t = \frac{54.3 - 50}{6.2/\sqrt{16}} = \frac{4.3}{6.2/4} = \frac{4.3}{1.55} \approx 2.77$$

8. **Chi-Square Test:**
With 5 degrees of freedom, the critical value at $\alpha = 0.05$ is 11.07.
Since $10.21 < 11.07$, it is not significant at the 0.05 level.

9. **Definition of CLT:**
The CLT states that as sample size increases, the sampling distribution
of the sample mean becomes approximately normal, regardless of population shape.

10. **Minimum Sample Size:**

$$\begin{aligned} n &= \left(\frac{z^* \cdot \sigma}{E}\right)^2 \\ &= \left(\frac{1.96 \cdot 10}{2}\right)^2 \\ &= (9.8)^2 \\ &= 96.04 \\ &\Rightarrow n = 97 \end{aligned}$$

11. **Standard Error for Difference in Proportions:**

    Calculate the sample proportions:

$$\hat{p}_1 = \frac{38}{80} = 0.475, \quad \hat{p}_2 = \frac{52}{100} = 0.52$$

Now compute the standard error step-by-step:

$$\begin{aligned} SE &= \sqrt{\frac{0.475 \times 0.525}{80} + \frac{0.52 \times 0.48}{100}} \\ &= \sqrt{\frac{0.249375}{80} + \frac{0.2496}{100}} \\ &= \sqrt{0.003117 + 0.002496} \\ &= \sqrt{0.005613} \\ &\approx 0.0749 \end{aligned}$$

12. **Standard Error for Difference in**

Proportions:

We are given:

$$\hat{p}_1 = \frac{38}{80} = 0.475, \quad \hat{p}_2 = \frac{52}{100} = 0.52$$

Now compute the standard error step-by-step:

$$SE = \sqrt{\frac{0.475 \cdot 0.525}{80} + \frac{0.52 \cdot 0.48}{100}}$$

$$= \sqrt{\frac{0.249375}{80} + \frac{0.2496}{100}}$$

$$= \sqrt{0.0031172 + 0.002496}$$

$$= \sqrt{0.0056132}$$

$$\approx 0.0749$$

Answer Key to Chapter 10 Exercises

1. **CI for Mean (σ Known)**
   Given: $\bar{x} = 14.5$, $\sigma = 3.2$, $n = 64$, $z^* = 1.96$ for 95% CI

$$SE = \frac{\sigma}{\sqrt{n}} = \frac{3.2}{\sqrt{64}} = \frac{3.2}{8} = 0.4$$

$$\text{CI} = 14.5 \pm 1.96 \times 0.4 = 14.5 \pm 0.784$$

$$\boxed{(13.716,\ 15.284)}$$

2. **CI for Proportion**

Given: $\hat{p} = \frac{280}{500} = 0.56$, $z^* = 2.576$ for 99% CI

$$SE = \sqrt{\frac{0.56(1 - 0.56)}{500}}$$

$$= \sqrt{\frac{0.2464}{500}}$$

$$= \sqrt{0.0004928} \approx 0.0222$$

$$\text{CI} = 0.56 \pm 2.576 \times 0.0222 = 0.56 \pm 0.0572$$

$$\boxed{(0.5028,\ 0.6172)}$$

3. **CI for Mean (σ Unknown)**
   Given: $\bar{x} = 890$, $s = 120$, $n = 40$, $df = 39$
   $t^* \approx 1.685$ for 90% CI and 39 df

$$SE = \frac{120}{\sqrt{40}} = \frac{120}{6.3246} \approx 18.97$$

$$\text{CI} = 890 \pm 1.685 \times 18.97 \approx 890 \pm 31.94$$

$$\boxed{(858.06,\ 921.94)}$$

4. **CI for Difference in Means**
   Given: $\bar{x}_1 = 6.1$, $s_1 = 0.4$, $n_1 = 25$; $\bar{x}_2 = 6.3$, $s_2 = 0.5$, $n_2 = 30$
   Assume unequal variances, use $t^* \approx 2.009$

(approx.) for 95% CI

$$SE = \sqrt{\frac{0.4^2}{25} + \frac{0.5^2}{30}}$$

$$= \sqrt{\frac{0.16}{25} + \frac{0.25}{30}}$$

$$= \sqrt{0.0064 + 0.0083}$$

$$= \sqrt{0.0147} \approx 0.1212$$

$$\text{CI} = (6.1 - 6.3) \pm 2.009 \times 0.1212 = -0.2 \pm 0.2433$$

$$\boxed{(-0.4433,\ 0.0433)}$$

5. **CI for Proportion & Interpretation**
   Given: $\hat{p} = \frac{108}{200} = 0.54$, $z^* = 1.96$

$$SE = \sqrt{\frac{0.54(1 - 0.54)}{200}}$$

$$= \sqrt{\frac{0.2484}{200}}$$

$$= \sqrt{0.001242} \approx 0.0352$$

$$\text{CI} = 0.54 \pm 1.96 \times 0.0352 = 0.54 \pm 0.069$$

$$\boxed{(0.471,\ 0.609)}$$

Conclusion: Since 0.60 is just barely inside the upper limit, the evidence does not strongly support the sociologist's claim at the 95% level.

Answer Key to Chapter 11 Exercises

1. **One-Sample t-Test:**

$$H_0 : \mu = 30, \quad H_1 : \mu > 30$$

$$t = \frac{32.4 - 30}{6.3/\sqrt{40}} = \frac{2.4}{0.9965} \approx 2.41$$

Degrees of freedom: $df = 39$ Critical t at $\alpha = 0.05$ (one-tailed): ≈ 1.685 Since $2.41 > 1.685$, we reject H_0. There is sufficient evidence to support the claim.

2. **One-Proportion z-Test:**

$$H_0 : p = 0.60, \quad H_1 : p \neq 0.60$$

$$\hat{p} = \frac{78}{150} = 0.52$$

$$SE = \sqrt{\frac{0.6 \times 0.4}{150}} = \sqrt{\frac{0.24}{150}} = \sqrt{0.0016} = 0.04$$

$$z = \frac{0.52 - 0.60}{0.04} = -2.00$$

Critical z at $\alpha = 0.01$ (two-tailed): ± 2.576 Since -2.00 is not beyond ± 2.576, we fail to reject H_0. No significant difference detected.

3. **Two-Sample t-Test (Unequal Variances):**

$$H_0 : \mu_1 = \mu_2, \quad H_1 : \mu_1 \neq \mu_2$$

$$\begin{aligned} SE &= \sqrt{\frac{8^2}{25} + \frac{10^2}{30}} \\ &= \sqrt{\frac{64}{25} + \frac{100}{30}} \\ &= \sqrt{2.56 + 3.3333} \\ &= \sqrt{5.8933} \\ &\approx 2.4276 \end{aligned}$$

$$t = \frac{81 - 76}{0.9194} \approx 5.44$$

Degrees of freedom (Welch's approximation): ≈ 50 Critical t at $\alpha = 0.10$ (two-tailed): $\approx \pm 1.676$ Since $5.44 > 1.676$, reject H_0. Significant difference in teaching methods.

4. **Two-Proportion z-Test:**

$$H_0 : p_1 = p_2, \quad H_1 : p_1 \neq p_2$$

$$\hat{p}_1 = \frac{48}{200} = 0.24, \quad \hat{p}_2 = \frac{60}{180} = 0.333$$

$$p_c = \frac{48 + 60}{200 + 180} = \frac{108}{380} \approx 0.2842$$

$$\begin{aligned} SE &= \sqrt{0.2842 \times (1 - 0.2842)\left(\frac{1}{200} + \frac{1}{180}\right)} \\ &= \sqrt{0.2842 \times 0.7158 \times 0.0111} \\ &= \sqrt{0.2037 \times 0.0111} \\ &= \sqrt{0.00226} \\ &\approx 0.0475 \end{aligned}$$

$$z = \frac{0.24 - 0.333}{0.0475} \approx -1.96$$

Critical z at $\alpha = 0.05$ (two-tailed): ± 1.96 Since $z = -1.96$, we are at the cutoff. It is a borderline case; conclusion may depend on rounding or strictness—commonly, we fail to reject H_0 if not beyond.

Answer Key to Chapter 12 Exercises

Matched Pair Testing:

$$d = \{4, 3, 3, 4, 1\}, \quad \bar{d} = 3.0, \quad s_d \approx 1.22, \quad n = 5$$

$$t = \frac{3.0 - 0}{1.22/\sqrt{5}} = \frac{3.0}{0.546} \approx 5.49$$

$$df = 4, \quad t_{\text{crit}} \approx 2.776 \Rightarrow \text{Reject } H_0$$

There is significant evidence that the training program improved endurance.

Exercise 2:

$$H_0 : \mu_1 = \mu_2 = \mu_3, \quad F = \frac{MS_B}{MS_W} = \frac{12.71}{4.2} \approx 3.03$$

With $df_1 = 2, df_2 = 12$, $F_{crit} \approx 3.89$. Since $F < F_{crit}$, fail to reject H_0. No significant difference.

Exercise 3:

$$\bar{X}_A = 50, \quad \bar{X}_B = 57.6, \quad \bar{X}_C = 44.4, \quad GM = 50.67$$

$$\begin{aligned} F &\approx 21.07 \\ F_{\text{crit}} &\approx 3.89 \\ &\Rightarrow \text{Reject } H_0 \\ &\Rightarrow \text{Fertilizer brand matters.} \end{aligned}$$

Exercise 4:

$$MS_B = \frac{75.2}{2} = 37.6, \quad df_W = 14 - 2 = 12$$

$$MS_W = \frac{144.8}{12} = 12.07, \quad F = \frac{37.6}{12.07} \approx 3.12$$

$$\text{Total SS} = 220.0$$

Exercise 5:

$$HSD = 3.96 \cdot \sqrt{\frac{6.2}{6}} \approx 3.96 \cdot 1.017 = 4.03$$

$$|\bar{X}_1 - \bar{X}_4| = 13.4 - 11.6 = 1.8 < 4.03 \Rightarrow \text{Not significant}$$

Significant pairs: 2 vs. 4, 3 vs. 4

Exercise 6:

$$\begin{aligned} S &= \sqrt{\frac{(4-1)(3.10)(6.2)}{2} \cdot \left(\frac{1}{6}+\frac{1}{6}\right)} \\ &= \sqrt{\frac{(3)(3.10)(6.2)}{2} \cdot \left(\frac{2}{6}\right)} \\ &= \sqrt{(9.3)(6.2) \cdot \left(\frac{1}{3}\right)} \\ &= \sqrt{57.66 \cdot \frac{1}{3}} \\ &= \sqrt{19.22} \\ &\approx 4.38 \end{aligned}$$

$$|\bar{X}_1 - \bar{X}_4| = 1.8 < 4.38 \Rightarrow \text{Not significant}$$

Answer Key to Chapter 13 Exercises

1. **Scatterplot and Correlation Interpretation**
   (a) The scatterplot shows a strong upward trend, suggesting a positive linear relationship between advertising and revenue.
   (b) The Pearson correlation coefficient $r \approx 0.986$, indicating a very strong positive correlation.
   (c) The strength is very strong, and the direction is positive — as advertising increases, revenue tends to increase as well.

2. **Simple Linear Regression Calculation**

   (a) The least-squares regression line is:

   $$\hat{y} = 2.074x + 0.045$$

   (b) Interpretation: For each additional \$100 spent on advertising, weekly revenue increases by about \$2,074. The intercept of 0.045 means if advertising spending were \$0, revenue would be approximately \$45 (in \$1,000s) — though extrapolating to 0 may not be meaningful.

   (c) If \$700 is spent on advertising ($x = 7$):

   $$\hat{y} = 2.074(7) + 0.045 = 14.518$$

   Predicted weekly revenue is approximately \$14,518.

3. **Assessing Model Appropriateness**

   (a) An $R^2 = 0.11$ means only 11% of the variability in performance scores is explained by hours of training.

   (b) No, the p-value of 0.24 exceeds typical significance levels (e.g., 0.05), so there is insufficient evidence to conclude a statistically significant relationship.

   (c) Possible reasons: (1) the relationship may be nonlinear, or (2) other confounding variables like job role or prior experience may be better predictors.

4. **Residual Plot Analysis**

   (a) A U-shape pattern indicates the residuals are not randomly scattered — the model underpredicts at the ends and overpredicts in the middle.

   (b) No, this suggests a nonlinear relationship, and the linear model is not appropriate.

   (c) Try fitting a quadratic regression or transforming the variables (e.g., log-rent or square-footage squared).

Bibliography

[1] Benjamin S. Bloom. Learning for mastery. *Evaluation Comment*, 1(2), 1968.

[2] John B. Carroll. A model of school learning. *Teachers College Record*, 64(8):723–733, 1963.

[3] I. Bernard Cohen. *The Newtonian Revolution*. Cambridge University Press, 2010.

[4] Tobias Dantzig. *Number: The Language of Science*. Free Press, 1954.

[5] Jean de Heinzelin. The ishango bone: An example of arithmetic in early africa. *Scientific American*, 196(6):108–117, 1957.

[6] Gary B. Deason. *A Mirror for the Eye: Early Enlightenment Science and the Invention of the Modern Mind*. University of Chicago Press, 2018.

[7] Graham Flegg. *Numbers: Their History and Meaning*. David & Charles, 1983.

[8] Edward Grant. *The Foundations of Modern Science in the Middle Ages: Their Religious,*

Institutional and Intellectual Contexts. Cambridge University Press, 1996.

[9] Ian Hacking. *The Emergence of Probability: A Philosophical Study of Early Ideas about Probability, Induction and Statistical Inference.* Cambridge University Press, 1975.

[10] Andrew Hodges. *Alan Turing: The Enigma.* Princeton University Press, 2012.

[11] Georges Ifrah. *The Universal History of Numbers: From Prehistory to the Invention of the Computer.* John Wiley & Sons, 2000.

[12] Fred S. Keller. "good-bye teacher...". *Journal of Applied Behavior Analysis*, 1(1):79–89, 1968.

[13] Alexander Marshack. *The Roots of Civilization: The Cognitive Beginnings of Man's First Art, Symbol and Notation.* McGraw-Hill, 1972.

[14] Seyyed Hossein Nasr. *Science and Civilization in Islam.* Harvard University Press, 2006.

[15] George Saliba. *Islamic Science and the Making of the European Renaissance.* MIT Press, 2007.

[16] Hugh Small. *Florence Nightingale: Avenging Angel.* Constable & Robinson, 2010.

[17] Doron Swade. *The Difference Engine: Charles Babbage and the Quest to Build the First Computer.* Viking Penguin, 2000.

[18] Betty Alexandra Toole. *Ada, the Enchantress of Numbers: Prophet of the Computer Age.* Strawberry Press, 2002.